PHYSICS
ENERGY OPTIONS

REVISED NUFFIELD ADVANCED SCIENCE
Published for the Nuffield–Chelsea Curriculum Trust
by Longman Group Limited

Longman Group Limited
Longman House, Burnt Mill, Harlow, Essex CM20 2JE, England
and Associated Companies throughout the World

First published 1986
Copyright in this format © 1986 The Nuffield–Chelsea Curriculum Trust.
The Trust is grateful to the original copyright owners of the articles included in this
book for their permission to reproduce this edited version.

Design and art direction by Ivan Dodd
Illustrations by Oxford Illustrators Limited

Filmset in Times Roman and Univers
Made and printed in Great Britain
by Scotprint Limited, Musselburgh, Scotland

ISBN 0 582 35424 2

———

Cover
Detail from 'Great Waves off Kanagawa' by Katsushai Hokusai (1760–1849).

CONTENTS

Introduction *page iv*

THE AGE OF FUSION: ON THE BRINK
Sam Walters *1*

BIOMASS FOR ENERGY: FUELS NOW AND IN THE FUTURE
Professor D. O. Hall *6*

OIL FROM ROCK
Sam Walters *12*

ISOTOPES IN GEOTHERMAL ENERGY EXPLORATION
E. Barbier, M. Fanelli, and R. Gonfiantini *17*

HEAT PUMPS PENETRATE THE INDUSTRIAL MARKETS
Arnold Witt *22*

FUEL FROM REFUSE
A. Porteous *27*

BRAZIL'S ITAIPÚ DAM
Joel Fagenbaum *32*

WAVE POWER: THE STORY SO FAR
B. M. Count, R. Fry, and J. H. Haskell *38*

RENEWABLE ENERGY: THE SEVERN BARRAGE
A. C. Baker *47*

THERMAL INSULATION OF BUILDINGS
E. A. Raynham *52*

STORAGE OPTIONS FOR HARNESSING WIND ENERGY
R. Ramakumar *55*

INTRODUCTION

This Reader is designed to support Unit G, 'Energy sources', of the Revised Nuffield A-level physics course. In part of that Unit individual students study one or more 'energy options', such as nuclear fusion, biomass, wind energy, conservation, and so on. A companion book to this one, *Energy sources: data, references, and readings*, gives some data on fuel supply and use, offers some advice about locating suitable material for study, and has an annotated bibliography of some 330 items.

For some of the suggested energy options there is a wide range of suitable references, many of which are readily available to schools and colleges, but material about other energy options is less easy to find. This book contains a collection of articles related to these energy options, in order to provide you with an easier start than would otherwise be possible. But the articles here will not, on their own, provide sufficient information for an effective study of any of the options.

In order to fit as many articles as possible into the available space the articles have been shortened, but in general they have not been rewritten: thus they still reflect the diverse styles of their authors. The articles were written for a variety of audiences – not for students studying for A-level – and so you may find things that you do not understand at first reading, or which require you to look elsewhere for clarification; some of the units and abbreviations may also be unfamiliar. This is part of the task involved in reading and reporting back to your colleagues on an energy option, or any other topic. Also bear in mind that conditions and state of research may have changed since the articles were written.

In terms of the energy-abundant world that would be ushered in, the achievement of controlled thermonuclear fusion ranks with the invention of fire. But controlled fusion is a new kind of fire; it is the ignition of a small star on Earth. Like nuclear fission, nuclear fusion releases enormous energy from small amounts of matter. Indeed, every form of life that has ever been present on Earth owes its existence ultimately to the fusion reactions that led to the formation of the planets. Our Sun is itself a giant reactor that converts 667.5 million tonnes of hydrogen every second into 663 million tonnes of helium; the difference of 4.5 million tonnes of solar mass becomes energy that pervades the entire Solar System, extending 5 billion miles (8 billion kilometres) into space.

The age of fusion: on the brink

SAM WALTERS
Staff Editor, *Mechanical Engineering*

A staggering potential

Here on Earth, the controlled fusion reaction of two or more isotopes of hydrogen into helium would solve the world energy problem for millennia at a tolerable fraction of the environmental damage caused by existing sources of energy. A few facts sum up fusion's potential. Hydrogen has three isotopes, of mass numbers 1, 2, and 3, known respectively as hydrogen, with the chemical symbol H, deuterium, with the symbol D, and tritium, with the symbol T. At the present level of research, the deuterium–tritium (D–T) reaction appears the most promising. Considering the amount of hydrogen in the oceans, deuterium can be considered superabundant (one atom to each 6700 atoms of ordinary light hydrogen). One cubic metre of water contains about 10^{25} atoms of deuterium, having a mass of 34.4 grams and a potential fusion energy of 7.94×10^{12} J. This is equivalent to the heat of combustion of 300 tonnes of coal or 1500 barrels (bbl) of crude oil. Since a cubic kilometre of seawater contains 10^9 m^3, its fuel equivalent is 300 billion tonnes of coal or 1500 billion bbl of crude oil. As for tritium, there is sufficient lithium (a source of tritium) in the United States alone to ensure, via the D–T reaction, an energy content more than fivefold that inhering in the World's fossil fuels (table 1).

Bottling a star

For fusion to occur, matter must first exist in the plasma state (so called by physicist Irving Langmuir in 1929 to describe its amorphous properties). Fusion takes place in the Sun because its enormous mass provides the gravitational force to squeeze and heat its substance to about 15 million °C. A combination of high pressure and high temperature continuously generates plasma by stripping atoms of their electrons and then driving their nuclei together with tremendous force. The result is a continuous flow of fusion energy.

The mass of the Earth, however, is so small in comparison with that of the Sun that the gravitational pressures needed for creating and confining a plasma long enough for fusion to occur are simply

Table 1
Estimated future sources of energy/Q or 10^{18} BTU ($\approx 10^{21}$ joules)

Source	Short term or known	Total future US supply
Gas or oil	0.6	2.8
Shale oil	0.9	9.6
Coal	5.2	34.0
Nuclear fission	1.3 (present reactors)	70.0* (breeder reactors)
Solar	A renewable resource	
Geothermal, wind, bioconversion	Limited to at best a few per cent of our total usage	
Fusion	250 (lithium)	16 000 000 000 (deuterium)

US energy use rate approximately 0.1 Q per year
* US uranium to $66 kg^{-1}

unattainable. Therefore we must enormously increase the factor of temperature and seek some means to confine the resulting plasma to effect a fusion reaction. The basic idea, then, of controlled fusion is the same today as it was when it was first conceived more than 25 years ago, preceding the explosion of the first hydrogen bomb. Take light elements, such as deuterium and tritium gas, and heat them to at least 100 million °C. This produces a plasma that can be confined either by its own inertia or, since the resulting plasma is composed of electrically charged particles, by magnetic fields (figure 1).

But there's the rub. Assuming that a deuterium–tritium plasma can be heated to a temperature of some 100 million °C, what kind of magnetic field would be needed to control the plasma? The difficulty does not lie in the very high temperature because, at the low gas densities in a fusion reactor (10^{14} particles cm^{-3}), the total energy content of the plasma would be insufficient to cause any serious damage to a containing vessel if the plasma were to come in contact with it. The plasma would, in fact, cool so rapidly that within a few minutes one could touch the walls without danger. The challenge is to find a magnetic configuration that will confine the volatile plasma fuel long enough, and a means of heating it quickly enough, so that the nuclei will interact before escape is possible to the container walls, where cooling inhibits fusion.

Magnetic bottles

In order to grasp the basics of the problem, let us look at the motion of plasma particles in a magnetic field. The presence of a magnetic field in a plasma causes the electrically charged particles to move in a particular manner. For example, in the absence of a magnetic field, the charged particles in an assembly (or plasma) held in a cylindrical vessel will move in straight lines in random directions and will quickly strike the walls. Now suppose a uniform (homogeneous) magnetic field is applied. The plasma particles will now be compelled to follow helical (*i.e.* corkscrew or spiral) paths about an axis lying inside and

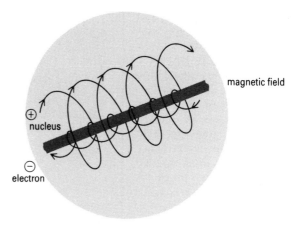

(a) Magnetic confinement

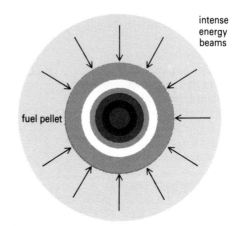

(b) Inertial confinement

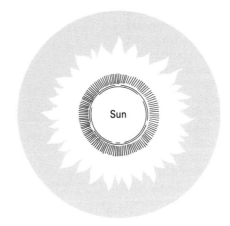

(c) Gravitational confinement

Figure 1
The three routes to fusion through confinement of a plasma.

parallel to the tube walls. The particles, therefore, all have two components of motion. The first is an orbiting motion perpendicular to the magnetic field lines, and the second a motion parallel to the field lines. This should keep the particles from touching the walls, a fusion requirement. But such a simple system provides no confinement in directions parallel to the tube axis.

The problem of the ends

Two answers arose to this 'problem of the ends'. Two general types of bottles were created: linear (open) and toroidal (closed) (figure 2). Open types generally rely on the so-called mirror effect. Here, intense magnetic-field regions (the mirrors), located at both ends of the confinement region, repel helically moving particles. This system thus traps charged particles between the ends and reflects them back and forth long enough to permit fusion reactions.

Closed systems bend the open tube into a circular shape, forming a torus, or doughnut-shaped figure. The particles then move freely along the magnetic field lines, which are all contained within the torus. To escape, the particles have to cross the field lines, a relatively slow process.

In principle, various magnetic configurations based on these two major approaches should provide adequate confinement for thermonuclear reactions, but the plasma simply refused to be confined long enough for sufficient fusions to occur regardless of the shape or strength of the magnetic bottle. Unstable gross motions and fine-scale turbulences encouraged either rapid expulsion of particles from the magnetic field or a somewhat slower, but still unacceptable, diffusion out of the field. The problem seemed unsolvable ... until 1958.

'Tokamania'

In 1958, a rather prosaic event took place – the 2nd International Conference on the Peaceful Uses of Atomic Energy. It was the

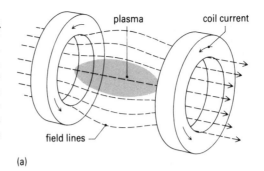

(a)

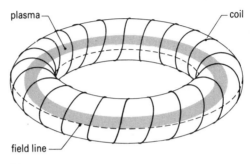

(b)

Figure 2
Two types of magnetic bottle.
(a) Open (linear), with the plasma moving to and fro along the field lines.
(b) Closed (toroidal), with the plasma moving along the helical field lines, confined within the torus.

first international meeting on this theme following the decision of the Eisenhower Administration to declassify the entire controlled-fusion programme. As it turned out, both the Russians and the Americans were following similar tracks.

Among the documents released at this meeting was a paper written in 1950 by Andrei D. Sakharov and Igor Tamm, which discussed an updated Russian version of a magnetic bottle called a tokamak. (The name tokamak is derived from the initial letters, to-ka-mak, of the Russian words meaning 'toroidal', 'chamber', and 'current', respectively.) This machine is a hollow, doughnut-shaped vessel in which the lines of force of a toroidal magnetic field are produced inside the vessel by electric currents passing through coils wrapped around the minor circumference of the torus. The Russians, in the early versions of this machine, did indeed confine the plasma – but not for long. As in the magnetic bottles designed in the United States and elsewhere, the hot particles crossed the field lines and crashed into the cold walls of the doughnut. But, in the 1950 paper, Sakharov and Tamm found the key to the confinement problem. They proposed that researchers make use of the fact that a heated plasma is an excellent conductor of electricity. At a temperature of 100 million °C it has a conductivity, in fact, about 30 times greater than that of copper. So they suggested that an electric current be induced along the plasma column, flowing the long way around the torus. This would create a second magnetic field, the so-called poloidal magnetic field, perpendicular to the first. This second field, much weaker than the first (the ratio of their field strengths is roughly 15:1), gives rise to helical field lines, which form a set of nested magnetic surfaces. In this configuration, the charged particles have closed orbits that do not depart from a given magnetic surface by more than their radius of gyration in the poloidal field.

As neatly expressed by Jeremy Bernstein, a physics professor at the Stevens Institute of Technology, in an article in *The New York Times* (Jan. 3, 1982): 'If you were to lay a doughnut flat on a counter top and slice down through it, the lines of the second magnetic field – the so-called poloidal field – would lie in the plane of the vertical slice; whereas the toroidal field lines go around in a horizontal circle parallel to the counter top. The combination of the two fields, the horizontal and the vertical, forces the plasma particles into so-called helical orbits – that is, into a spiral path around the hole in the doughnut, staying within the magnetic fields and not touching the cold walls'. (See figure 3.)

By 1969, their machine, the Russians claimed, heated plasma to 10 million °C

while maintaining densities of 30 000 billion particles per cm^3 for 1/50 second. The Soviet claim met with some scepticism, since plasma measurements are notoriously difficult to make and to interpret. A British team, however, from the Culham Laboratory in Abingdon, later confirmed that the temperature was indeed at the level claimed by the Soviets.

Following proof of the Russian claim, the dream of controlled fusion seemed attainable for the first time. A shock wave, known as 'tokamania', went through the international community. Within a short time the tokamak became the 'mainline approach' to fusion throughout the World. Today, numerous operational tokamak experiments are under way.

The fusion–power balance

What is the fundamental requirement for a meaningful release of fusion energy in a reactor? A fusion reaction can be sustained when the energy released is greater than that required to heat and sustain the temperature of the plasma.

To do this, it is necessary to maintain the plasma in a stable condition at a minimum temperature of at least 100 million °C, such that the product of ion density and confinement time ($\eta\tau$) is about 10^{14} s cm^{-3}. This requirement is known as the 'Lawson criterion' and describes the minimum condition for attaining the break-even fusion reaction. At higher values (3×10^{14} s cm^{-3}), the high-temperature helium nuclei from fusion reactions provide sufficient heating to be self-sustaining; therefore, no external heat-

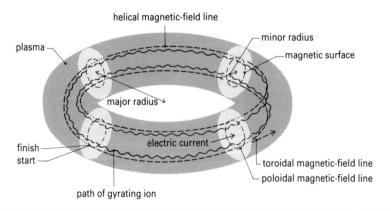

Figure 3
Principle of a tokamak, showing the toroidal and poloidal magnetic fields which combine to keep the plasma in a confined helical path.

The age of fusion

ing is required. This is referred to as the 'ignition condition'. Figure 4 shows the progress towards the goal of an ignited plasma through the graph of achieved and projected results on contours of constant thermonuclear energy gain.

A number of reactions other than the D–T fusion are of potential use; the most feasible are shown in figure 5, giving the energies released in each reaction. However, because of the relatively low temperature required and the high reaction energy, a favourable energy gain is far easier to achieve with D–T fusion than with any of the other reactions shown in the figure. When greater temperatures, densities, and confinement times can be achieved, other reactions have major advantages; for example, they may:

1 Eliminate the need for tritium breeding. (Tritium is radioactive, with a half-life of $12\frac{1}{2}$ years.)
2 Use only naturally occurring fuel (*e.g.* deuterium).
3 Allow the design of reactors that would produce electricity directly (*i.e.* without the use of a thermal heat cycle).

As has been shown, the energy resource provided by nuclear fusion is enormous. However, twenty-six years after Sakharov and Tamm's paper was made public the design problems involved in producing an operational fusion reactor have still not been solved. Although much experimental work is taking place, looking into various types of reactor, it is not known when, or whether, nuclear fusion will be able to help with the World's energy needs.

(This article first appeared, in a slightly longer version, in *Mechanical Engineering*, July 1982. It is reproduced with kind permission.)

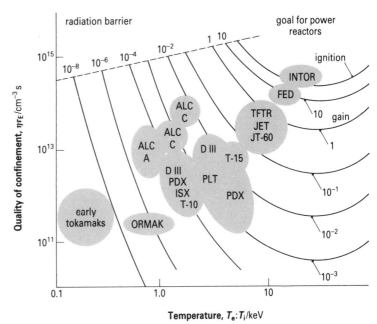

Figure 4
The progress towards the goal of an operational fusion reactor. Reactors must combine high temperature and confinement in order to reach ignition.

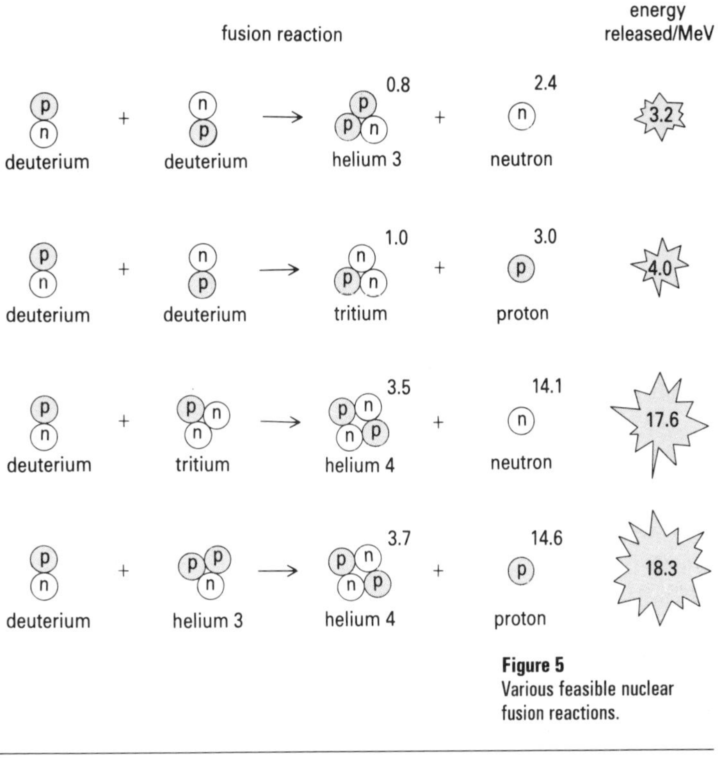

Figure 5
Various feasible nuclear fusion reactions.

Biomass for energy: fuels now and in the future

PROFESSOR D. O. HALL

School of Biological Sciences, King's College, London

Biomass is a jargon term used in the context of energy for a range of products which have been derived from photosynthesis. You will recognize the products as waste from urban areas and from forestry and agricultural processes; specifically grown crops, sugar crops, hydrocarbon plants and oils, and also aquatic plants such as water weeds and algae. Thus, everything which has been derived from the process of photosynthesis is a potential source of energy. We are talking essentially about a solar energy conversion system, since this is what photosynthesis is. One of the disadvantages of coming from the United Kingdom and talking about solar energy is that nobody believes that we have any. Let me allay your fears on this and indicate that the mean radiation in the UK is 100 watts per m^2, averaged over 24 hours for the whole year. The mean in the United States is 185 watts per m^2 and the mean in Australia is 200 watts per m^2. The difference between ourselves and these sunny countries is only a factor of two – it is not ten or a hundredfold as most people imagine.

The problem with solar radiation is that it is diffuse and it is intermittent, so that if we are going to use it we have to capture a diffuse source of energy and we need to store it – this is what plants solved a long time ago.

If you look inside any green leaf you will see cells with chloroplasts containing the chlorophyll, where photosynthesis occurs. The process of photosynthesis embodies the two most important reactions in

life. The first one is the water-splitting reaction which evolves oxygen as a by-product from H_2O; all life depends on this reaction. The second reaction is the fixation of carbon dioxide to organic compounds. All of our food and our fuel is derived from this CO_2 fixation from the atmosphere. It should be appreciated that these two reactions require inputs of light in two sequential reactions. When looking at an energy process we need to have some understanding of what the efficiency of this process will be; we need to look at the efficiency over the entire cycle of the system, and, in the process of photosynthesis, we mean incoming solar radiation converted to a stored end-product. Table 1 indicates the losses incurred, starting from incoming solar radiation to eventual stored energy products. Most people agree that the practical maximum efficiency of photosynthesis is between five and six per cent. It might not seem very good, but I should remind you that this represents stored energy. Photovoltaic solar cells can operate at 15 % efficiency, but this is instantaneous conversion from solar radiation to electricity and does not include storage. One must always look at the entire cycle from incoming radiation to final storage over the whole year.

The photosynthetic efficiency will determine the biomass dry weight yields.

Table 1
Photosynthetic efficiency and energy losses

	Available light energy
At sea level	100 %
50 % loss as a result of 400–700 nm light being the photosynthetically usable wavelengths	50 %
20 % loss due to reflection, inactive absorption, and transmission by leaves	40 %
77 % loss representing quantum efficiency requirements for CO_2 fixation in 680 nm light (assuming 10 quanta/CO_2)*, and remembering that the energy content of 575 nm red light is the radiation peak of visible light	9.2 %
40 % loss due to respiration	5.5 %
	(Overall photosynthetic efficiency)

* If the minimum quantum requirement is 8 quanta/CO_2, then this loss factor becomes 72 % instead of 77 %, giving the final photosynthetic efficiency of 6.7 % instead of 5.5 %.

Table 2 shows that in the UK, with 100 watts per m^2, a good potato crop growing at one per cent efficiency will yield between twenty and twenty-five tonnes dry weight per hectare per annum. Obviously, if we can grow and adapt plants to increase their photosynthetic efficiency, the dry weight yields will increase and, of course, alter the economics of the crop. One of the very interesting areas of research is to try to understand what the limiting factors are in photosynthetic efficiency in plants both for agriculture and for biomass energy.

There is another aspect of photosynthesis that we should appreciate. This is that the health of our biosphere and atmosphere is totally dependent on the process of photosynthesis. Every three hundred years all the CO_2 in the atmosphere is cycled through plants; every two thousand years all the oxygen, and every two million years all the water. Thus, three key ingredients in our atmosphere and biosphere are dependent on cycling through the process of photosynthesis.

World energy use

How much photosynthesis actually takes place on the Earth? Table 3 (page 8) shows that the World's total annual use of energy is only one tenth of the annual photosynthetic energy storage; i.e., photosynthesis already stores ten times as much energy as the World needs. The problem is getting it to the people who need it. Secondly, the energy content of stored biomass on the Earth's surface today, which is 90 % in trees, is equivalent to our proven fossil fuel reserves. In other words the energy content of the trees is equivalent to the commercially extractable oil, coal, and gas. Thirdly, during the Carboniferous era (340–280 million years before the present) quite large quantities of photosynthetic products were stored, but in fact they only represent one hundred years of net photosynthesis. The overall photosynthetic efficiency during the Carboniferous era was less than 0.000 2 %. Fourthly, we must remember the problem of CO_2-cycling in the atmosphere. Many people are rightly concerned about the problem of the build-up of CO_2 in the atmosphere if we con-

Table 2
Average-to-good annual yields of dry matter production

	Yield/tonnes hectare^{-1} yr^{-1}	Yield/ g m^{-2} day^{-1}	Photosynthetic efficiency (per cent of total radiation)
Tropical			
Napier grass	88	24	1.6
Sugarcane	66	18	1.2
Reed swamp	59	16	1.1
Annual crops	30	–	–
Perennial crops	60–80	–	–
Rain forest	35–50	–	–
Temperate (Europe)			
Perennial crops	29	8	1.0
Annual crops	22	6	0.8
Grassland	22	6	0.8
Evergreen forest	22	6	0.8
Deciduous forest	15	4	0.6
Savanna	11	3	–
Desert	1	0.3	0.02

tinue to burn fossil fuels. Again, it is a problem of cycling between two or three 'pools' of carbon. The amount of carbon stored in the biomass is approximately the same as the amount of atmospheric carbon and the same as the carbon stored as CO_2 in the ocean surface layers; there are three equivalent pools. The problem is how the CO_2 is distributed between these pools, and how fast it equilibrates into the deep ocean layers.

Developing countries

There is another very important aspect of biomass energy use in the World. This is the fact that at least half the World's inhabitants are primarily dependent on biomass as their main source of energy. This fact has only been widely appreciated in the last few years – it has been called the 'fuelwood crisis' or the 'second energy crisis'. There is no doubt that half the people in the World have a much more serious energy problem than the rest of us, who are primarily concerned about the price of petrol and whether we should turn the heat up or down.

If you examine the statistics you will see that about 15 % of the World's annual fuel supplies are currently derived from biomass. The average person in the rural areas of the developing world uses the equivalent of about one tonne of wood per year. This is mainly for cooking and heating but also for small-scale industry, agriculture, food processing, and so on. This 15% of the World's energy use represents the equivalent of twenty million barrels of oil per day, which is ten times as much oil as we get out of the North Sea, and equivalent to the total United States oil use. I don't think that many people realized the importance of this, since the statistics were not available to show the significance, and the consequences of biomass overuse were not readily evident. In fact, until a few years ago, World energy supply statistics listed biomass at 3, 4, or 5 per cent, if at all.

It is now known that about half of all the trees cut down in the world today are used for cooking and heating. The problem of deforestation, with its consequent flooding, desertification, and agricultural problems, is not solely due to overcutting of trees for cooking and heating. There are obviously other factors involved, such as commercial and illegal cutting, absence of replanting, and so on, but we do not have space to discuss these here.

Recently a few very good papers have been published on studies in southern and eastern Africa which show that in an average family of six or seven, one person's sole job is to collect firewood, and they will often have to walk great distances, which of course has other deleterious consequences. In urban environments, households may spend up to 40 % of their income on fuelwood and charcoal.

Another aspect, which has been highlighted in Tanzania, is the curing of tobacco; for each hectare of tobacco you need to burn the wood from one hectare of savanna woodland. There are many examples of the fact that it is not only domestic fuelwood use but also agricultural, urban, and small-scale industrial uses which are having very adverse long-term effects.

Serious attempts are being made by a number of national and international groups to try to reverse this problem of deforestation by vigorously promoting reforestation, village fuelwood lots or community forestry, agroforestry, and so on. One study published by the US National Academy of Sciences in 1981 is a long-overdue manual of tree species especially suited for fuelwood in the humid tropics, the arid tropics, and temperate regions. Another study, published by the International Council for Research in Agroforestry in Nairobi, promotes the concept of agroforestry, from which one can derive both food and fuel from the reforestation schemes.

The World Bank concluded in 1981 that if one were to reverse the deforestation problem, one would need to spend 6.75 billion dollars over the next five years in order to reforest fifty million hectares. There is little hope that this will happen, but it was realistically thought to be needed. There are a number of reasons

Table 3
Fossil fuel reserves and resources, biomass production, and CO_2 balances

Proven reserves	Tonnes coal equivalent	Energy equivalent
Coal	5×10^{11}	
Oil	2×10^{11}	
Gas	1×10^{11}	
	8×10^{11}	25×10^{21} J
Estimated resources		
Coal	85×10^{11}	
Oil	5×10^{11}	
Gas	3×10^{11}	
Unconventional gas and oil	20×10^{11}	
	113×10^{11}	300×10^{21} J
	Tonnes carbon equivalent	
Fossil fuels used so far (1976)	2×10^{11}	6×10^{21} J
World's annual energy use	5×10^{9} (from fossil fuels)	3×10^{20} J
Annual photosynthesis		
(a) net primary production	8×10^{10}	3×10^{21} J
(b) cultivated land only	0.4×10^{10}	
Stored in biomass		
(a) total (90 % in trees)	8×10^{11}	20×10^{21} J
(b) cultivated land only (standing mass)	0.06×10^{11}	
Atmospheric CO_2	7×10^{11}	
CO_2 in ocean surface layers	6×10^{11}	
Soil organic matter	$10-30 \times 10^{11}$	
Ocean organic matter	17×10^{11}	

These data, although imprecise, show that (a) the World's annual use of energy is only 10 % of the annual photosynthetic energy storage, (b) stored biomass on the Earth's surface at present is equivalent to the proven fossil fuel reserves, (c) the total stored as fossil carbon fuel only represents about 100 years of net photosynthesis, and (d) the amount of carbon stored in biomass is approximately the same as the atmospheric carbon (CO_2) and the carbon as CO_2 in the ocean surface layers.

why this won't be possible, but I shall mention only one here. Due to the very low status that foresters have in developing (and also developed) countries not only are good foresters relatively scarce, but so also are the employment opportunities for them. It is all very well promulgating reforestation schemes to help solve the energy crisis in various parts of the World, but unless you have the people on the ground with the experience and the knowledge, there is no way that these schemes can be implemented.

Biomass attributes

Biomass as a source of energy has problems and it has advantages. Like every other energy source, it is not the universal panacea. Some advantages and disadvantages are listed in table 4.

I wish to emphasize one advantage that is of particular interest to me, namely the large biological and engineering development potential which is available for biomass. Presently we are using knowledge and experience which have been static for many years; the efficiency of production and use of biomass as a source of energy has not progressed in the way that agricultural yields for food have increased. Thus, there is an undoubted potential to increase biomass yields. The most obvious problem that comes to mind is the competition for land between biomass and food production. Existing agricultural, forestry, and social practices are also a hindrance to promoting biomass as a source of energy whether in a developing or a developed country. However, development of practices using biomass production as part of an overall agricultural strategy could go a long way towards solving this.

Vegetable oils

What about ways of deriving liquid fuel? It has been known for a long time that we can use all types of vegetable oils in diesel engines. In fact, in 1911 Rudolf Diesel (1858–1913) advocated the use of vegetable oils in his engines in agricultural regions of the World, and he predicted that it would become important in the future. Studies in Zimbabwe, South

Table 4
Some advantages and problems foreseen in biomass for energy schemes

Advantages	Problems
1. Stores energy.	1. Land and water use competition.
2. Renewable.	2. Land areas required.
3. Versatile conversion and products with high energy content.	3. Supply uncertainty in initial phases.
4. Dependent on technology already available with minimum capital input; available to all income levels.	4. Costs often uncertain.
5. Can be developed with present manpower and material resources.	5. Fertilizer, soil, and water requirements.
6. Large biological and engineering development potential.	6. Existing agricultural, forestry, and social practices.
7. Creates employment and develops skills.	7. Bulky resource; transport and storage can be a problem.
8. Reasonably priced in many instances.	8. Subject to climatic variability.
9. Ecologically inoffensive and safe.	
10. Does not increase atmospheric CO_2.	

Africa, Australia, Brazil, the Philippines, the United States, Austria, Germany, and other countries have shown that in the sunny countries if a maize farmer devoted 10 % of his land area to growing sunflowers or peanuts he could run all the diesel-powered machines that he uses in his farming operation. It is not generally advised that these vegetable oils be used pure; probably a blend of from 10 to 30 % is preferable. The question is whether to spend money on refining oil or to use it unrefined – if used unrefined, you have to make sure that oil filters and jets are cleaned frequently.

There is also a lot of interesting work going on about the esterification of sunflower oil as methyl or ethyl esters. The esterified oil has fuel properties very close to those of diesel and the esterification can be done on the farm. The Brazilians are devoting most of their effort to extraction of oil from peanuts and also looking carefully at palm oil as a 6 % blend into diesel.

Hydrocarbon plants

What about using plants directly to produce gasoline? Such proposals have been around for quite some time, with the main proponent being Calvin of the University of California. He advocates growing the plant *Euphorbia lathyrus* for the extraction of hydrocarbons which have molecular weights very close to that of gasoline. There are also large trials going on, mostly in Arizona, financed by the Diamond Shamrock Corporation, to establish whether this is economically viable or not. Unfortunately, the initial claims of high yields were not substantiated, but the recent studies show 10 barrels (1.5 tonnes) of oil per hectare per annum can be produced. The question is whether these yields are sustainable in arid environments. The answer is as yet unknown. However, there are at least half a dozen more trials in various parts of the World looking into whether this is economically viable.

There is another requirement from oil, and that is the manufacture of synthetic rubber. Guayule, which grows naturally in northern Mexico and the southern USA, can be used as a source of rubber with properties which are indistinguishable from that of the rubber tree. In 1910 Rockefeller and Vanderbilt made a fortune supplying half the World's rubber from such guayule bushes. There is now a pilot plant producing one tonne a day in Saltillo, Mexico, and a fifty tonne a day plant is being proposed in northern Mexico.

An alga, *Botryococcus braunii*, has been shown in Australia to yield 70 % of its extract as a hydrocarbon liquid closely resembling crude oil. This has led to the work and ideas in France of immobilizing these algae in solid matrices such as alginates and polyurethane and using a flowthrough system to produce hydrocarbons. A green alga called *Dunaliella*, discovered in the Dead Sea, Israel, produces glycerol, beta-carotene, and also protein. The alga does not have a cell wall and it grows in these very high salt concentrations; thus, to compensate for the high salt externally, it produces glycerol internally. A recent publication showed that if glycerol, betacarotene, and protein are produced this can be an economically viable system.

Biogas

There are many energy problems which cannot be solved by technology, and I just want to mention one of them. One can see cow-dung patties beautifully stacked in high piles outside New Delhi; cow-dung sellers gather the cow-dung in the surrounding area and then sell it in the cities where it is used as fuel. This has serious consequences to agriculture in the areas surrounding such cities, since one does not want to remove the manure fertilizer from the soil. There are, however, means of obtaining energy from cow-dung, and at the same time producing fertilizer as a by-product. The process involves fermentation in biogas digesters, which produces methane, and has been known for a long time. It has been realized that biogas digesters could be very useful in many parts of the World.

There is, however, no universal prescription in advocating biogas as an energy source in developing countries. In southern China they have a biogas digester which is a single family unit being built at about the rate of one million a year in Szechwan Province – seven million have been constructed so far. They are cheap, not very efficient, but they do work. In India they have about one hundred thousand biogas digesters, which are efficient, with steel domes and concrete bases, but they require a minimum of about five cows in order for them to work efficiently. Unfortunately, there is not a great percentage of families in India with five cows. Thus, ideally such biogas digesters would be best suited to a community or a village; but there are social problems in putting this type of system into a village and this can be the case in many parts of the World.

Photochemistry and photobiology

I wish to conclude with an area of research in which we are involved in our laboratory, and with which I am very much concerned, namely how we might use photochemical and photobiological systems, not in ten years, but maybe fifteen or twenty years hence. We know that plants fix CO_2 and that they require soil, fertilizers, and water, and that they must be protected from all types of predators. Hence, is it possible to short circuit the plant and construct systems which can directly fix CO_2, or split water, or fix nitrogen, without the intervention of the plant? Thanks to Calvin and Benson in California, we know the path of photosynthetic carbon fixation from CO_2 in the atmosphere to sugars, carbohydrates, fats, and so on in the plant. Unfortunately, we do not know how the crucial initial step works. The enzyme ribulosebisphosphatecarboxylase, which fixes all the CO_2 in the atmosphere (which has gone into our oil, coal, gas, and currently goes into our food and into our trees) is the most abundant enzyme on Earth. Half the protein in the lettuce leaf that you eat comprises this single enzyme. Unfortunately, we do not know how it works, even though we are learning quite a lot about it.

Can we mimic the CO_2-fixation process? There are two reports from Israel and Japan which describe the photochemical fixation of CO_2 to methanol, formic acid, and formaldehyde. These are both photoelectrochemical systems which work only in ultra-violet light. However, it is early days and these are the first reports on fixing CO_2 from the atmosphere into fuels.

What about the splitting of water? Jules Verne said in 1874: 'I believe that water will one day be employed as fuel, that hydrogen and oxygen, which constitute it, used singly or together, will furnish an inexhaustible source of heat and light'. Unfortunately he did not tell us how to do it and we still do not know exactly how to do it. There have been reports from a number of laboratories on photochemical systems using colloidal particles incorporating titanium and ruthenium, which can split water and produce hydrogen and oxygen. There are still experimental problems with such systems, but they are the first reports of purely chemical systems that look as if they might function.

The only system which actually does function in visible light on a continuous basis is the biological system. In such biosystems we want to use the water-splitting reaction, throw away all the CO_2-fixation components, add in enzymes from bacteria, shine light on the system, and produce hydrogen gas plus oxygen as a by-product. All very well, but unfortunately it is a biological system and it eventually dies. We and other groups have tried to immobilize the system so that it can continuously produce hydrogen and oxygen, because if the system is ever going to be practically useful it has to work for years and not hours. A system immobilized in alginate on a steel grid produces hydrogen and oxygen in the light, but it dies after about twelve hours. We have been able to substitute the enzymes using platinum or synthetic iron–sulphur catalysts. What we have not been able to do is to substitute for the water-splitting reaction which is the crux of the whole problem.

Can we mimic, or can we construct, a system which will split water using visible light so that the electrons from water can be used to produce high-energy compounds like fixed carbon, hydrogen, or ammonia? At present the answer is no. What we can do quite positively is to use waste materials like organic compounds or sulphur (with which you don't have to split water) to give continuous production of hydrogen gas upon illumination. In our laboratory we have had such a system running for fifteen days continuously producing pure hydrogen from compounds such as ascorbic acid, formate, and so on. But we have still not solved the ultimate problem of splitting water.

Conclusion

What I have tried to say is that the process of photosynthesis is really quite marvellous. We start with water, with light, and use membranes containing chlorophyll to produce an energy-rich compound which is usually an iron–sulphur catalyst. Normally the energy in that compound is used to fix carbon dioxide into a whole range of organic compounds. But we can also use this ultimately to fix nitrogen, reduce sulphur, reduce oxygen, and I have just shown you that you can use it to produce hydrogen. Plants and their derived systems can do nearly anything.

(This article first appeared, in a slightly longer version, in the *Journal of the Royal Society of Arts*, Volume **CXXX**, Number 5312, July 1982. It is reproduced with kind permission.)

Oil from rock

SAM WALTERS
Staff editor, *Mechanical Engineering*

There's an old saying that you can't get blood out of a stone. But you can get oil out of a rock – if it's oil shale. One American pioneer, the story goes, learned this the hard way. He built his hearth from the rock – and when he lit a fire in his new fireplace, his hearth caught fire and his cabin burned down.

Oil shale is actually laminated marlstone (see figure 1), containing a solid substance called kerogen, a carbonaceous material between asphalt, crude oil, and coal. Over the ages nature never provided the necessary heat or pressure to turn kerogen into natural crude oil. Man, however, can finish the job by heating the rock to a temperature between 450 and 600 °C. At that temperature the kerogen turns to vapour and separates from the rock. The vapour cools to a liquid resembling crude oil, which, after upgrading, can be used as a refinery feedstock.

The largest known oil shale deposits in the US were left behind by a long-vanished freshwater lake that once covered most of Colorado, Utah, and Wyoming. This lake bed geologic area, the Green River formation, covering some 16 000 square miles (41 000 km²), holds some of the richest deposits of oil shale in the World – up to 1500 ft (460 m) in thickness.

Commercial development is feasible only for yields of at least 25 US gallons (115 litres) of oil per ton of feed shale. Western shales, yielding between 25 and 40 US gallons/ton (115 and 185 litres tonne⁻¹ respectively), meet this standard, whereas eastern shales averaging less than 15 US gallons/ton (69 litres tonne⁻¹) are considered poor candidates for commercial development. They may be better suited, as fuel prices escalate, for gasification than they are for conversion to liquids.

There are two basic shale oil extraction techniques: surface retorting and modified *in situ* retorting. The first technique involves physically mining the shale. In one process, the shale is mined normally in surface or in 'room-and-pillar' mines (underground mines with ceilings supported by pillars of unmined rock). The shale is then crushed, preheated, and fed into huge retorts in which hot ceramic balls, ½ in (13 mm) in diameter, are mixed with it in a rotating drum. The heat from the balls converts the kerogen into vaporized shale oil. The oil vapour is then cooled and condensed and sent through the upgrading plant where nitrogen, sulphur, and other

Figure 1
An example of an oil shale.
A Shell photograph

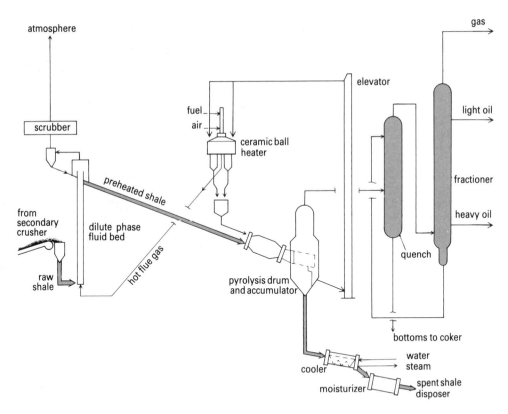

atmosphere

gas

scrubber

fuel

air

elevator

ceramic ball
heater

light oil

preheated shale

fractioner

from
secondary
crusher

dilute phase
fluid bed

hot flue gas

heavy oil

raw
shale

quench

pyrolysis drum
and accumulator

bottoms to coker

water
steam

cooler

moisturizer

spent shale
disposer

Figure 2
The surface retorting process
for oil shale.

impurities are removed. Shale oil leaves the site by pipeline. (See figure 2.) The hot, spent shale, a dark-grey, gravel-like substance, now expanded in volume, or 'puffed', is then moistened (a process that uses most of the water required by shale projects, about 3 gal/bbl), and disposed of in a site where it will be compacted, contoured, and revegetated.

The *in situ*, or underground, heating process eliminates the usual mining. Explosives are detonated underground to reduce the unmined shale to rubble in a previously partially evacuated space. Steam and air are then passed through the rubble to create a combustion zone that moves through the bed. Kerogen, released by the heat, turns to gaseous hydrocarbons, which cool to liquids. These drain to the floor and are pumped to the surface for processing.

Western oil shale reserves alone, it is estimated, hold 1800 billion bbl, a figure far surpassing the petroleum reserves of the entire Arab world. More than 600 billion of these barrels can be recovered with present technology, an impressive

figure when one realizes that only some 100 billion bbl of conventional petroleum have been brought to the surface in the USA since the start of oil exploration. However, there are formidable obstacles (outlined below), the least of which is the technology of the processes themselves that must be overcome before more than a small fraction of this potential can be realized. A more realistic figure published by various government and industry studies suggests an output of 400 000 to 600 000 bbl day^{-1} of crude shale oil by 1990.

Three major concerns

Three major considerations could limit the large-scale development of shale oil: availability of water, environmental factors, and socio-economic considerations.

Availability of water

A 50 000 bbl day^{-1} shale oil plant would process about 30 million tonnes of raw shale per year and would require the disposal of about 14 million m^3 of compacted spent shale annually. A lot of water is required – water to obtain and process the crude shale oil, and additional water to cool the spent shale and establish new vegetation on top of it. This may not always be available at the source of the base fuel and must be brought in. One oil company estimates, according to the American Petroleum Institute (API), that water from Colorado's local sources would permit development of a shale oil industry producing up to 2 million bbl day^{-1} by the late 1990s. Beyond that time, however, to support continued production, water may have to be piped in from far-off reservoirs.

Environmental factors

Crude shale oil, says Alan Schriesheim, General Manager of the Engineering Technology Department of Exxon Research and Development Co., contains a large amount of nitrogenous compounds and significant amounts of arsenic. These and various other elements present in oil shale are potential pollutants. In addition, the spent shale contains salts that are potentially leachable, as well as organic pyrolytic products. How these products of oil shale processing will be treated under the US Toxic Substances Control Act and how wastes will be categorized under the US Resource Conservation and Recovery Act of 1976 could have a big effect on development.

There is also the problem of controlling the release of CO_2 to the atmosphere. Retorting of oil shales at above about 600 °C may release more CO_2 through decomposition of carbonate minerals than the amount of CO_2 subsequently generated by burning the oil produced.

Socio-economic considerations

The aesthetic aspects of huge projects of the type projected for the Green River formation are also important. A Tosco official has estimated that Colony's mine in Parachute, Colorado, in which the traditional room-and-pillar method will be used to produce 'rooms' 18 × 18 m, big enough to hold a small apartment house,

will be bored for a distance of 14 500 km, many times as long as the New York subway system. Today, the site 'is an awesome spectacle, with giant earthmoving equipment looking like specks against the sides of huge brown, grey, and white cliffs of shale'. As one Exxon official put it at a recent symposium on synthetic fuels, a topographic metamorphosis will occur in which 'canyons are turned into mountains'.

Energy officials are confident, however, that all problems are solvable. B. M. Louks, the Electric Power Research Institute's projects manager for engineering, expects shale oil to become price-competitive with petroleum in the near future – before the end of the 1980s.

Coal to liquid

Coal is the most abundant available energy resource both in the US and worldwide. US coal production – about 780 million tons in 1979 – is expected to triple by the year 2000, with about 30 % being used in the manufacture of synthetic fuels. Coal is basically a rock of an organic sedimentary heterogeneous material, characterized by differences in the kinds of original plant materials (type), degree of metamorphism (rank), and the amounts and types of impurities present.

The classification of coals by rank is what most people have some familiarity with: coal can be anthracite, bituminous, or lignite, with subvariations of these ranks. This classification is according to a fixed carbon and calorific value (Btus per pound), calculated for carbon free of minerals. Wood was the dominant plant material in typical peat swamps; therefore, typical coals consist predominantly of the wood-derived vitrinite (one of the organic compounds of coal).

The four roads to coal-oil

The conversion of coal to liquid fuel requires not only a reduction in molecular weight, but also a substantial increase in the hydrogen-to-carbon ratio. There are four general methods for the liquefaction of coal: pyrolysis, direct liquefaction, indirect liquefaction, and chemical synthesis.

Pyrolysis

Pyrolysis, or destructive distillation, breaks down the heavy molecular structures of coal, which contain high ratios of carbon atoms. This produces liquids and tars having much smaller molecular structures and much lower ratios of carbon to hydrogen. Its major product, however, is a carbonaceous char, which is almost entirely carbon – a nonvolatile residue. Any process, to be commercially viable, must make use of this char through combustion or other gaseous reactions. Recent research on pyrolysis and hydropyrolysis – pyrolysis in the presence of hydrogen – has involved fluidized bed techniques and other processes. Investigations are also focusing on the use of pyrolysis as the first step in a more complex process in which the residual char is upgraded by gasification, using steam and/or oxygen, and the liquids and tars are upgraded by hydrogenation and other refining operations.

Direct liquefaction

Several advanced methods of coal conversion take this route – the addition of hydrogen to make liquids with higher hydrogen-to-carbon ratios. Hydrogenation was first used with coal in 1913 by the German chemist, Friedrich Bergius (1884–1949). Three variations of the Bergius process are now being tested in pilot and demonstration plants in the US. All involve mixing powdered coal with a solvent and treating it with hydrogen at high temperatures and moderately high pressures. Two of the processes – Gulf's SRC (solvent refined coal) and Exxon's EDS – use a solvent to 'donate' the hydrogen to the coal. The third, the H–coal process, adds hydrogen directly. A catalyst is used to speed the chemical reaction.

Indirect liquefaction

South Africa is the World's largest commercial producer of synthetic fuel from coal. Her three SASOL plants, when fully operational (two are completed), will turn out close to $100\,000\,\text{bbl}\,\text{day}^{-1}$ at a cost well below that of imported oil. The technique used is a two-stage or 'indirect' type: coal is converted into carbon monoxide and hydrogen, which is then recombined into a wide range of liquids, hydrocarbons, and others. This recombination is the basis of the German-developed Fischer–Tropsch method developed in Germany during World War II to supply oil, which could not be imported. (Figures 3 and 4.)

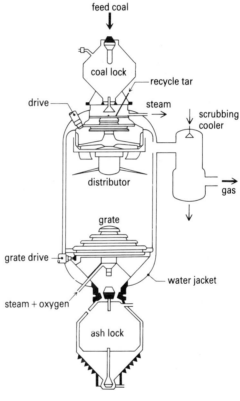

Figure 3
The Lurgi gasifier, 84 of which will be employed at the SASOL complex.

Figure 4
The Fischer–Tropsch process used in the indirect liquefaction process of transforming coal into oil.

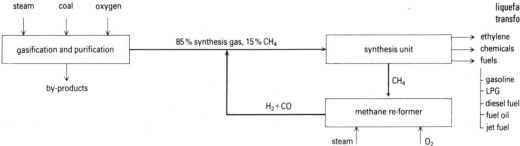

Chemical synthesis

Methanol is the first of the series of alcohols. For 100 years it came from the pyrolysis of wood. Now it is made from syngas, $CO + 2H_2$, produced from any source of carbon: coal, lignite, peat, natural gas, petroleum distillate, or residue that is too far from the market or too low in heating value, or perhaps has too high a content of combined water, sulphur, and/or ash to have any other value.

The production of methanol from syngas involves rather simple chemical processes. Syngas is produced either by the reforming of methane with steam or by the partial oxidation of the natural gas or other base fuel with pure oxygen. In either case there results an impure mixture of one or both carbon oxides and hydrogen. This 'synthesis', or syngas, is purified, all sulphur is removed – as brimstone – down to parts per billion, and the ratio of these gases is adjusted. One mole of carbon monoxide combines with two moles of hydrogen at elevated temperature and pressure in beds of catalysts to give the methanol product. Methanol, as it comes from the reactor of Wentworth Bros. Inc., is ready for immediate sale with no distillation or other purification required. In addition to 97% methanol, the liquid contains some higher alcohols and 1% water. These by-products aid the combustion. Methanol is a true 'synthetic' fuel, because it is a combination or synthesis of the simple molecular gases formed by breaking down the heavy molecules of the coal.

As in the case of the hydrocarbon synfuels – from coal, lignite, tar sands, oil shales, and so on – there are heavy plant costs. (In synfuel plants the cost of capital may be from one-half to two-thirds of the total product costs.) The production, purification, and other processing of the synthesis gases, and particularly the preliminary separation from air of the large tonnage of oxygen – 25 tons min^{-1} in the largest plants so far designed – require a vast assemblage of mechanical handling equipment, pumps, compressors, furnaces, reactors, and equipment for gas separation, heat interchanging, and recovery of waste heat.

However, there are some significant pluses. One is the water requirement. In a Wentworth plant designed for processing a low-grade lignite at a place in the western US where water is scarce, there will be no water required. Water, chemically combined in the lignite as mined, will actually be produced in excess of that required for the reaction and boiler feed. This excess will be sold to supply local needs. This contrasts dramatically with the needs of all other syncrude production processes which are dependent on large quantities of good (high-quality) water. Such water is often not available at the source of the base fuel.

Of equal significance is the use of energy in this plant, now being engineered as the largest one of a long series of such plants extending back to the mid-1920s. All process steam and electric power for the entire operation, including that of the tremendous plant for separating oxygen from air, comes from the recovery of heat otherwise lost in the methanol synthesis, because of improvements in catalysts and their usage.

End of an era

We now face a major energy transition, as we enter an era of 'rock burning' that may last 100 years or longer before the technology is ready for a nigh inexhaustable resource such as fusion – deriving its energy from 'sea burning' – or the ultimate resource, solar energy.

One of the foremost problems facing humanity today, asserts Hubbert, is how to make the cultural adjustments appropriate for a stable high-technology and high-energy, but essentially non-growing, society of the future. 'If we succeed,' he says, 'we could achieve a state of well-being, which could provide an environment for the flowering of a great civilization. Should we fail, the consequences are not pleasant to contemplate'.

(This article first appeared, in a slightly longer version, in *Mechanical Engineering*, February 1982. It is reproduced with kind permission.)

The future of geothermal energy presents questions which are not easy to answer. It is certain that up to now only the more accessible geothermal fields have been exploited and also that in the near future it can become important for the industrial and social development of many regions in developing countries. However, the particular attraction of geothermal energy is that its ultimate source is practically infinite: the heat of the Earth's interior. The challenge, therefore, is how to exploit even only the distant periphery of such an immense reservoir of abundant energy.

Isotopes in geothermal energy exploration

E. BARBIER and M. FANELLI
Istituto Internationale per le Ricerche Geotermiche, Pisa
R. GONFIANTINI
International Atomic Energy Agency, Vienna

Heat from the depths of the Earth

Since the 18th century, measurements in mines have shown that temperature increases with depth: in other words, that there is a continuous heat-flow from the Earth's interior to the surface. The average temperature gradient is 3°C per 100 m depth.

From the surface to the centre, the Earth is made up broadly of three layers: the crust, the mantle, and the metallic core. The crust, mainly composed of granites above and basalts beneath, has a thickness of about 35 km under the continents; under the oceans the crust is only 5 km thick, and consists mainly of basaltic rocks. The crust floats on top of the denser and more plastic materials of peridotitic type which constitute the mantle which has a thickness of 2900 km. The crust is split into six main plates which move relative to each other. (See figure 1.)

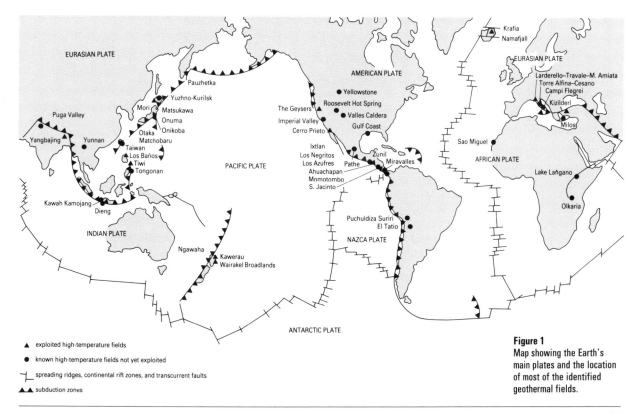

▲ exploited high-temperature fields

● known high-temperature fields not yet exploited

⊥ spreading ridges, continental rift zones, and transcurrent faults

▲▲ subduction zones

Figure 1
Map showing the Earth's main plates and the location of most of the identified geothermal fields.

Along the so-called mid-oceanic spreading ridges, the mantle materials ascend to the Earth's surface, determining the growth of adjacent plates. These plates, therefore, move away from each other at a rate ranging between 2 and 20 cm per year. In geological terms this is an enormous speed. To compensate for this growth there are other zones, called subduction zones, where one plate plunges below the adjacent plate into the mantle. These zones often correspond to oceanic trenches, bordered by island arches or by mountain ridges along the continental edges. Mid-oceanic spreading ridges and subduction zones tend to be the areas where the internal heat of the Earth reaches the surface, and where most volcanoes and most of the World's geothermal fields are located. (See figure 2.)

The geothermal fields

Geothermal fields are areas where the temperature of the groundwater is well above normal values and where the water can be exploited for various purposes, such as space heating and power production (above 150 °C). The heat sources are magmatic intrusions at depths of 7 to 15 km or, at places where the Earth's crust is thinner, the mantle itself. The heat is transferred to sub-surface regions firstly by conduction and then by groundwater convection. Impermeable rocks cover the permeable formations containing the hot water (called the geothermal reservoir), preventing or limiting the heat losses and maintaining the hot fluid under pressure. If boreholes are sunk to the geothermal reservoir, the hot fluid can be extracted and exploited. (See figure 3.)

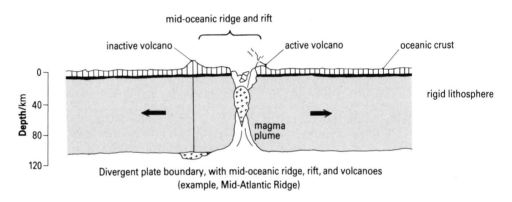

Divergent plate boundary, with mid-oceanic ridge, rift, and volcanoes
(example, Mid-Atlantic Ridge)

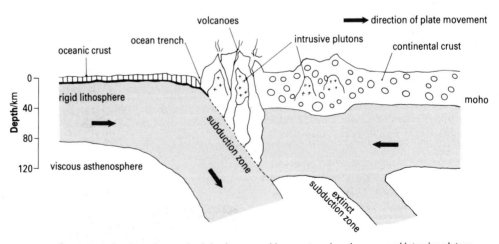

Convergent plate boundary, and subduction zone with ocean trench, volcanoes, and intrusive plutons
(example, Peru–Chile Trench)

Figure 2
Schematic representation of a mid-oceanic ridge and a subduction zone.

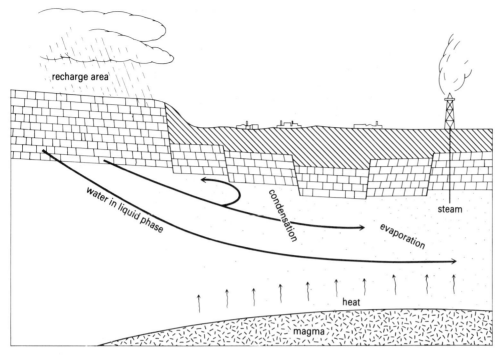

water in liquid phase

condensation

evaporation

steam

heat

magma

Figure 3
Schematic representation of a geothermal field.

When the conditions are particularly favourable, superheated dry steam can be produced with a temperature often exceeding 250 °C. This can be conveyed directly to turbines to move alternators and produce electricity. Such geothermal electricity is the cheapest of all the ways of producing electrical power, as shown in figure 5. Examples are the fields of Larderello and Monte Amiata (Italy) (figure 4), The Geysers (California, USA), and Matsukawa (Japan).

Less favourable, but still economically suitable for power production, are those geothermal fields delivering mixtures of steam and hot water. In these cases, the water has to be separated from the steam before the steam goes to the turbines. Examples of such fields are Wairakei (New Zealand), Cerro Prieto (Mexico), and other geothermal fields in Japan, in the Philippines, and elsewhere. The hot water generally carries large amounts of dissolved salts, which may give rise to problems of corrosion and incrustation.

When the temperature of the hot groundwater is only 50 to 100 °C, the water is mainly used for space heating (apartment blocks, greenhouses, and so on) as in the People's Republic of China,

France, Hungary, Iceland, the USSR, and other countries. A further use is for medicinal and bathing springs.

The geothermal plants so far in operation in the World have a total installed power of only about 2700 MW (see table 1), corresponding, for example, to about 2 % of the World capacity of nuclear power stations. To this power capacity,

Figure 4
A steam-producing well at Larderello at the moment of blow-out, when drilling reaches the geothermal reservoir.
International Atomic Energy Agency, Vienna

Figure 5
The interior of a geothermal power station at Larderello.
ENEL, Italy

however, one should add the non-electrical uses of geothermal energy. In addition, there are a number of already identified geothermal fields which it is planned to exploit during the 1980s. They are in the following countries: Chile, Greece, Guatemala, India, Nicaragua, and the West Indies. Geothermal exploration is being carried out in many other countries, and promising results have been obtained, especially in Argentina, Bolivia, Bulgaria, Colombia, Costa Rica, Djibouti, Ecuador, Ethiopia, Iran, Israel, Peru, Poland, Spain (Canary Islands), and Thailand. In many developing countries geothermal power production may in future meet an appreciable fraction of their energy needs.

In future, it is hoped to extract the heat of the Earth's interior from deeper and deeper formations, including deep, hot rocks which are compact and dry. In such a case two wells would be drilled and hydraulically inter-connected at depth through a network of artificially created fractures. Cold water would be injected in one well and hot water recovered from the other. Preliminary experiments at 3000 m depth done at Los Alamos, New Mexico, USA have shown the feasibility of such projects (see figure 6). The experiments are continuing in the USA, and others have been undertaken in the UK (Cornwall), to develop and refine the technology.

Table 1
Geothermal power plants in the World

Country	Installed power in 1982/ MW
China, People's Republic of	4
El Salvador	95
Iceland	41
Indonesia	30
Italy	440
Japan	174.5
Kenya	30
Mexico	180
New Zealand	202
Philippines	570
Portugal (Azores Is.)	3
Turkey	0.5
USA	936
USSR	11
Total	2717

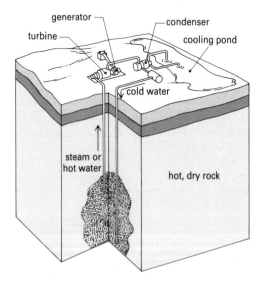

Figure 6
Schematic representation of an artificial geothermal field. The two wells are connected by an artificial network of fractures.

Geothermal energy

Natural isotopes as a tool in geothermal exploration

Water is the most abundant component of the geothermal fluid and is also the main energy carrier, since water has a high heat capacity and a high latent heat of vaporization. Other gaseous components accompanying the steam phase of a geothermal fluid are carbon dioxide (0.5 to 5 %); hydrogen sulphide, ammonia, nitrogen, methane, hydrogen (up to 2 or 3 % altogether); boric acid; and rare gases.

The first contribution that isotope techniques made in the understanding of geothermal fields was in demonstrating that hot water and steam derive from rainwater infiltrating from the outcrops of the geological formations constituting the geothermal reservoir. Although now it seems rather trivial, this hypothesis was far from being generally accepted about thirty years ago, before deuterium and oxygen-18 content measurements demonstrated it definitely. Before then, scientists believed that the geothermal steam was of juvenile origin, that it originated from the Earth's interior and was appearing for the first time on the Earth's surface.

Isotopic measurements, especially when carried out in parallel with conventional geochemical analyses, provide information on the characteristics of the geothermal fields: for example, on the mixing between waters of different temperatures and origin, on the underground flow patterns of the hot fluids, on the degree of interaction with the reservoir rocks, on the origin of the various fluid components, and so on. However, the most important use of isotope techniques in geothermal exploration is probably to estimate the deep temperature. This parameter is generally not easily accessible, but is essential in assessing the potential of geothermal fields. In particular, the use of isotopes is invaluable during the preliminary geochemical exploration of geothermal fields. When drillings have not yet been made, samples are collected from springs, geysers, and mofettes. Isotopes are also important when the boreholes do not reach a sufficient depth or are insufficient in number.

Isotopic geothermometers exploit the fact that the isotope distribution between two components in thermodynamic equilibrium depends only on temperature. The most commonly used isotopic geothermometers are those based on hydrogen isotopes in the systems methane/hydrogen gas and water/hydrogen gas, on oxygen isotopes in the system water/dissolved-sulphate, and on carbon isotopes in the system carbon dioxide/methane.

Geochemical thermometers have also been developed based on the silica content of the hot waters, on the relative concentrations of sodium, potassium, and calcium, or on the composition of the gas-steam phase. All these geothermometers, including the isotopic ones, are affected to differing extents by processes occurring during the ascent to the surface of the geothermal fluids, such as mixing processes, evaporation losses, interaction with the rocks encountered, partial re-equilibration at lower temperature, and so on. However, the parallel use of as many geothermometers as possible helps in evaluating correction terms and in attaining the right result.

Tritium has been used to identify modern water in geothermal fluids. As is known, atmospheric thermonuclear tests have injected large amounts of tritium into the hydrological cycle, especially in the years 1952 to 1962. Therefore, tritium can be used to evaluate mixing of shallow groundwater and to trace the arrival and the flow patterns of modern recharge from the edges of geothermal fields.

Recently, isotopes have been used to follow the fate of reinjected geothermal fluids. This is a new and interesting application which results from the attempts to reduce the environmental impact of the geothermal fluids: instead of releasing them into surface waters after they have been exploited, they are reinjected into the geothermal field and 'recycled'. Among other things, this may also increase the production of the field.

(This article first appeared, in a slightly longer version, in the *International Atomic Energy Agency Bulletin*, Volume **25**, Number 2, June 1983. It is reproduced with kind permission.)

Heat pumps penetrate the industrial markets

ARNOLD WITT
Electricity Council

In recent years well over 350 electrically driven heat pumps have been installed in the UK. Their installed load amounts to some 25 MW. As well as providing a correspondingly larger energy saving, other tangible production benefits have been achieved.

Today, more than ever before, industrial management is looking to maximize the return from all its resources, including capital, labour, raw materials, manufacturing space, and energy. That electrically driven heat pumps can fulfil a role in industry is demonstrated by the continually increasing numbers installed.

Principle

In a heat *engine* (such as a steam engine, internal combustion engine, or gas turbine) the energy flow is from hot source to cold sink, with an output of useful work. A heat *pump* may be thought of as a heat engine running in reverse. The most familiar example is a refrigerator, in which an input of energy (or work) is used to make heat flow from cold to hot.

Essentially the heat pump system consists of a compressor and two heat exchangers with appropriate valves. One of the heat exchangers (the evaporator) abstracts heat from a source at a relatively low temperature and the other (the condenser) gives off this heat, plus that required to drive the compressor, at a higher temperature.

Electric motors are almost universally used to drive the compressor. However, internal combustion engines firing on natural gas, LPG, or diesel oil may also be used. They also offer the potential to recover heat from their cooling water and exhaust systems. Overall capital, installation, maintenance, stand-by, and other operating costs of both types of heat pump have to be taken into account, as well as the energy cost savings.

Advantages

Heat pumps offer two main attractions. First, provided the temperature rise between the two heat exchangers is not too great, a heat pump will transfer more energy than is required to drive it. This relationship is described by the expression coefficient of performance (COP). The expression should not be used loosely because its value will depend not only on temperature differences, but on the hours run and the additional energy used by any necessary auxiliary plant.

The second advantage of the heat pump is its ability to recover heat from a low temperature source and make it available at a higher temperature.

Heat recovery

Heat pumps have been, and will continue to be, applied extensively in order to recover heat from low temperature sources, *e.g.* cooling water effluent.

Although individual circumstances will differ, the best prospects for installing a heat pump for heat recovery purposes will arise when:

1 Both the cooling and heating capabilities of the heat pump are employed – to make the best use of capital.

2 The heat pump is called upon to run for relatively long periods of time, *e.g.* it is used for process or process-related heating rather than seasonal, environmental heating.

3 Warm water heat sources are available. This enables higher use temperatures to be achieved together with economically attractive COPs.

In the food industry, in particular, studies by the Electricity Council* point to the considerable potential for heat pumps for heat recovery.

Certain sectors of the food and drink industries are heavy users of refrigeration. The notable sectors are meat, poultry, and fish processing, dairies, chocolate confectionery, fruit and vegetables, and the brewing industry. In these sectors, the heat rejected from the refrigeration plant, particularly if water-cooled, provides a vast input source of heat for heat pumps. Water leaving the condenser and entering the cooling tower would typically be at 28 °C.

Additionally, most of these same sectors have high demands for hot water for domestic hygiene and utensil and plant cleaning purposes: useful temperatures could be within the range 50–70 °C. Space and boiler-feed heating can also benefit from a supply of heated water.

In fact, the amount of heat rejected nationally from the refrigeration systems, in these food industry sectors, could comfortably provide their space heating and factory services heating demands. This is illustrated in table 1.

Further plus points suggest that heat pumps will play an increasing role in these sectors. The difference between source and use temperature is sufficiently small to achieve economically attractive COPs. The use temperature is within the technical capacity of modern plant – a number of equipment suppliers can supply standard plant. A range of design and contracting

*See WITT, J. A. *Electric heat pumps for heat recovery from food industry cooling water systems.* The Electricity Council, 30 Millbank, London SW1P 4RD.

Table 1
Comparison of services and steam usage, recovered heat and factory operating patterns

Sector	Heat rejection from refrigeration plant/ TJ yr^{-1}	Potential† heat recovery/ TJ yr^{-1}	Steam for space heating and hot water/ TJ yr^{-1}	Main shift pattern ($\times$ 8 hr)
Meat and fish	2420	2420	3020	1, 2, and 3
Milk	2850	2850		
Chocolate confectionery	1250	1250	580	3
Fruit and veg.	1280	1280	1500	2 (variable)
Total	7800	7800	5100	

† Assumes a COP of 3.5:1 and a diversity factor of 70 %

expertise exists across the UK through the Refrigeration and Unit Air Conditioning Group (RUAG) of the Heating, Ventilating, and Contractors' Association (HVCA). Heat pumps, too, will often make it possible to reduce the use of cooling towers, so saving on water, treatment, and effluent costs. Finally, many of the sectors operate on two or more shifts, so making the economic case much more attractive.

Examples of heat pump heat recovery installations were beginning to appear by the early 1980s and are continuing to increase.

Two electrically driven heat pumps installed at the Milk Marketing Board's Bamber Bridge bottling plant may lay claims to pointing the way. In this case the heat pumps, driven by motors totalling 171 kW, recover heat from a chiller cooling water circuit (27 °C) and from a water treatment plant (24 °C). The output of hot water, up to 32 therms per hour at 60 °C, is used for a variety of purposes, such as preheating boiler feed water, domestic hot water, and crate and vehicle cleaning.

At the meat products factory of Northern Foods Meat Group, Nottingham, a 151 kW heat pump installation is producing 537 kW of heat energy at an operating temperature of 65 °C. This installation, by Hilton Building Services (Nottingham), using a Carlyle 'Heat Machine', provides the factory's space heating requirements and preheats hot water for both cleaning and hygiene purposes. A back-up steam-heated calorifier is provided to cover the eventuality of there being no waste heat available from the product refrigeration plant's cooling water system – the usual source of heat for this heat pump system. Estimated payback on the capital investment of £33 500 in 1981 is between two and a half and five years.

Drying

Another area in which the heat pump is finding wide application is in drying. Electric heat pump dryers, sometimes called dehumidifiers or dehumidification dryers, can offer many real advantages over batch, fuel-fired, convection plant. This is particularly true for materials for which drying is dependent on the rate at which moisture migrates through the material as well as from its surface.

A heat pump drying system comprises a chamber, an internally or externally mounted heat pump, an air circulation system, auxiliary heating, and controls.

Heat pump dryers differ from convection drying systems in two important respects. Firstly, the latent heat of vaporization of water (or other solvent) is recycled. Consequently, apart from an initial warm-up period, any additional heat input will be minimal. Secondly, moisture (or solvent) is drained away rather than vented to the atmosphere. This allows the chamber to operate as a closed system for most of the cycle, and also achieves better control of drying. The principle of the operation is shown in figure 1.

Many applications of heat pumps for drying are to be found in the food, textiles, ceramics, and timber sectors.

Food
Drying at Smith Kendon, manufacturer of fruit pastilles and Juicy Jellies, was originally provided by dual-fired boilers using gas or oil. Average loads took about 42 hours to dry, but improved methods and increased production requirements were placing further demands on the drying facilities.

One drying chamber was converted to heat pump dehumidification as a pilot scheme in 1982, and the remaining two were converted in 1983. Each chamber houses three 4 kW electric heat pump dehumidifiers, which can maintain 30 % relative humidity at an air temperature between 30 °C and 60 °C to suit the product being dried. The dehumidification chamber is shown in figure 2.

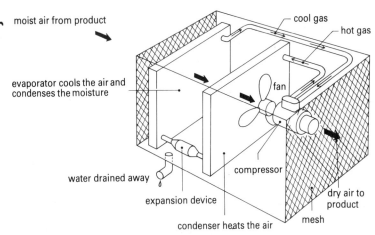

Figure 1
The operation of a heat pump dryer.

moist air from product · cool gas · hot gas · evaporator cools the air and condenses the moisture · fan · water drained away · compressor · expansion device · dry air to product · condenser heats the air · mesh

Figure 2
The heat pump dehumidification chamber at Smith Kendon.
The Electricity Council

Careful monitoring of the chambers under production conditions shows an 80% energy cost saving, which paid for the cost of the complete installation in 18 months. There is also a saving in production time of up to 60%, with a corresponding increase in throughput without requiring the corresponding increase in number of trays or drying space.

Textiles

Kangol Wear Ltd., headwear manufacturers, has changed its drying chambers from oil-fired steam heating to electric heat pump dehumidification (as shown in figure 3). Table 2 shows detailed comparisons between the two systems. The substantial energy cost savings have paid for the installation cost in 16 months.

Ceramics

At Henry Watson's Potteries the main products are terra cotta mouldings or castings, which have to have their moisture content of 12 to 25% on a dry weight basis (dwb) reduced to less than 0.5% before firing in an electric kiln. The original oil-fired systems often required periods of

Table 2
Textile drying

		Chamber I. Woollen berets	Chamber II. Hanks, cheeses, ribbons
Oil-fired system	**Cycle time**	5.75 h	5.75 h
	Oil usage	95.68 l (1004 kW h equivalent)	77.85 l (817 kW h equivalent)
	Electrical energy consumption	13.2 kW h	32.1 kW h
	Water removal	154 kg	54 kg
	Specific energy consumption	6.60 kW h kg^{-1}	15.72 kW h kg^{-1}
Heat pump dehumidifier system	**Cycle time**	4.5 h	4.7 h
	Heat pump capacity	15 kW	7.5 kW
	Energy consumption including fans	82.7 kW h	46.43 kW h
	Water removal	145 kg	54 kg
	Specific energy consumption	0.57 kW h kg^{-1}	0.85 kW h kg^{-1}
Comparison	**Energy usage reduction on site**	91%	94%
	Reduction in energy costs	70%	75%

Figure 3
The dehumidification chamber at Kangol Wear Ltd.
The Electricity Council

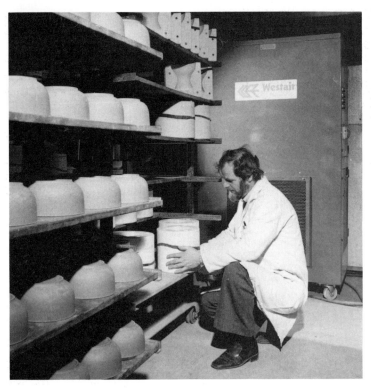

Figure 4
Heat pumps are used to
dry out ceramic pots at
Henry Watson's Potteries Ltd.
The Electricity Council

continuous operation of up to 17 hours. The moisture-laden air was vented to the atmosphere.

A free-standing, 7.5 kW electric heat pump dehumidifier was easily installed and reduced drying times typically to 12 hours. The reduction in energy costs was even greater at 45 %, since the heat pump not only recovers and recycles the sensible heat but also the latent heat from the moisture-laden air which is no longer vented. The cost of the system was recovered in about two years. In addition, the improved conditions in the drying chamber (figure 4) enable a wider variety of products to be dried at the same time. This, together with the reduction in energy costs, has enabled the firm to achieve high product quality, increased production, and competitive prices.

Timber
The use of an 80 °C heat pump dryer, developed by the Electricity Council Research Centre and Westair, leads to savings of 65 % in energy and 40 % in costs for drying timber, as shown in table 3. This machine can also be used for drying a variety of materials.

Future

The Electricity Council Research Centre is continuing its work on heat pumps, and new developments are now available to breweries and distilleries.

Because of the energy cost savings arising from the use of heat pumps, and because of the other important production cost saving benefits which they yield, the technology is gaining acceptance in many industrial sectors. A continuing and more widespread adoption can be forecast with confidence.

(This article first appeared in *Energy Management Focus*, Number 1, April 1985. It has since been shortened. It is reproduced with kind permission.)

Table 3
Comparison of timber drying in a conventional kiln and a heat pump dehumidifier kiln

Process	Kiln load/m^3	Moisture reduction (%)	Drying time/h	Energy consumption/ kW h m^{-3}	Energy cost/£ m^{-3}
Heat and vent – oil	21	27–14	93	309 (equivalent)	4.18
Dehumidifier	11.5	27–14	93	75	2.17

Fuel from refuse

A. PORTEOUS
Reader in Engineering Mechanics, Open University

The cost of conventional fossil fuels has risen substantially in recent years, and this is likely to continue as fossil fuels become scarcer. This article describes how refuse can be used to produce fuel, which can be used for heating instead of fossil fuel. Over the same period, traditional methods of disposal of refuse have also become more difficult and more expensive. Thus the use of 'refuse derived fuel' (RDF) has the potential for savings on two separate counts.

Conventional disposal methods

The cheapest method of disposing of refuse is to tip it into a large hole, conveniently situated close to the source of the refuse (the so-called landfill method). Very often there are problems due to smells, blow-away refuse, and contamination of water supplies. Even when such sites are available close to the urban centres which produce most refuse, they eventually become full and sites which are further away have to be used. This sometimes involves transferring the refuse from the local collection vehicles to barges or rail wagons, and can multiply the cost by four times. Even so about 50% of the refuse of London has to be dealt with in this way. Methods which could reduce this problem would clearly be attractive.

Burning refuse

Table 1 shows the proportions of the various constituents of typical municipal refuse as it is collected. This table shows a characteristic of western refuse, that it contains a high proportion of paper and other combustibles. It is possible to reclaim the paper but it is often not economical to do so, even if it is collected separately from the rest, and, in practice, once it is mixed with other refuse it is there to stay. A common solution has been simply to burn the refuse, and to dissipate the heating effect in the atmosphere. This is being increasingly rejected since the incinerators are expensive to build and maintain, and little or no income is derived from the process.

However, the potential of refuse as a fuel is shown by comparing the calorific values of crude refuse, paper within the refuse, and UK industrial coal (table 1). Since the UK produces about 20 million tonnes of domestic refuse per year, this is equivalent to about 6 million tonnes of industrial coal, which is about 5.5% of the UK consumption of coal in 1982.

A number of instances are documented of the direct use of refuse for heating, and perhaps also for the generation of electrical power. It is not economical to generate power from refuse in the UK because the power generated has to be sold to the Central Electricity Generating Board at a price determined by the CEGB. If a refuse-burning power station were to be built today, the cost of disposing of the refuse would be double the cost of transporting it to remote landfill sites, and much more than cheaper methods of disposal.

Table 1
The composition of refuse

Composition	% by weight (national average)	Average specific enthalpy as received*/ MJ kg^{-1}
Fine dust (20 mm)	19	9.6
Paper	30	14.6
Vegetable	21	6.7
Metals	9	nil
Glass	9	nil
Rag	3	1.6
Plastic	3	37
Unclassified (wood, leather, etc.)	6	17.6
Crude refuse (overall)	100	9–11
UK industrial coal (for comparison)	–	24.5–28

* Moisture content typically 20–30% by weight

The obvious method of burning refuse, to heat industrial processes or the domestic and commercial buildings near the incinerator, depends for its viability on a number of factors. Suitable industries or buildings must be located nearby, which is usually a matter of chance since it is not usually possible to plan from scratch. The demand for energy must be more or less constant, which can often be satisfied for industrial heating but the demand for domestic heating is heavily dependent on climatic factors. It has been possible to meet these requirements in some cases in Germany and the USA, and even sometimes to include generation of electricity, but the charges made for the energy are much more under the control of the seller than they are in the UK.

Refuse derived fuel (RDF)

The major advantages of RDF are that it can be used where it is needed rather than where it is made, and it can be stored, to be used when it is needed. The local authority would not need to provide expensive boiler plants, nor to find markets for the steam or power.

Producing and using RDF

Loose RDF
Loose RDF is useful where there is an installation which can use the fuel conveniently close to the refuse disposal plant. Figure 1 is a schematic diagram of the production of locse RDF for use in making cement.

The refuse is initially crushed into large pieces of between 100 and 200 mm. The smaller pieces are screened out, and subsequently passed through magnetic separators to extract the ferrous metals. The larger pieces of refuse are shredded a second time and passed through the screening process again. The refuse now consists of relatively small pieces, without any ferrous metal content, and is passed along to the cement kiln. In this particular example the refuse can be used to replace 10–20 % of the fuel value of the coal which is normally used to fire the kilns. In other situations the loose RDF can be used to replace up to

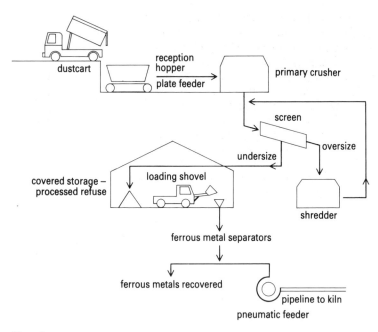

Figure 1
Loose RDF production system.
Blue Circle Group

half of the fuel value of the coal, which can result in substantial fuel-cost savings.

The screening process could involve no more than allowing the refuse to drop through suitably sized holes in a mesh, but air classifiers allow the paper content to be separated much more effectively (figure 2).

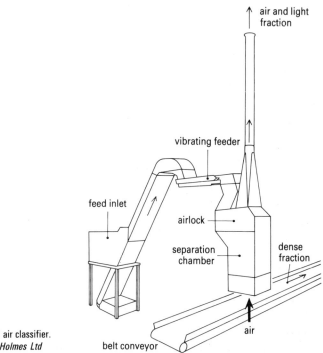

Figure 2
A vertical air classifier.
Peabody Holmes Ltd

The pulverized refuse is fed via an airlock into an air stream moving quickly upwards. By adjusting the feed rate, the particle size, and the air velocity, 90 % of the combustibles can be extracted from the refuse. The resultant material can be burnt as it stands or subjected to further shredding. Figure 3 shows a flowchart for an installation in Chicago, which is typical of many in the USA, which have been set up to take advantage of the economic advantages of this method compared with other methods of refuse disposal. However, many of the incinerator–boiler installations in the USA have suffered from severe boiler corrosion.

Pelleted RDF

Pelleted RDF is useful where there is no installation which could use the fuel conveniently close to the refuse disposal plant. Figure 4 shows RDF pellets which are suitable for being transported or stored before being used.

The very low sulphur content of RDF means that sulphur dioxide emissions to the environment may be greatly reduced, especially when it replaces high-sulphur coal, as in the USA. Nevertheless some sulphur dioxide is produced, together with chlorine and hydrogen chloride. Depending on the type of installation in which the RDF is being used, these gases are responsible for corrosion of any boiler tubes through which they may pass. Table 2 gives typical values of various properties of RDF and coal.

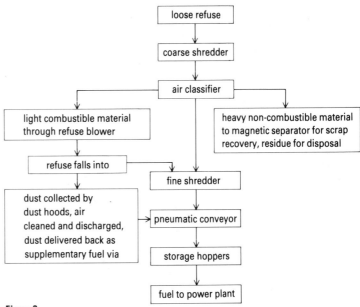

Figure 3
Diagrammatic representation of the operation of the City of Chicago southwest supplementary fuel processing facility.

Figure 4
20 mm diameter refuse derived fuel pellets.
Tyne and Wear County Council

Table 2
Comparison of the properties of RDF and coal

Fuel	Specific enthalpy/ MJ kg^{-1}	Moisture (%)	Ash (%)	Volatile matter (%)	Sulphur (%)
Pelleted RDF ex-Doncaster refuse	15.7	16.8	15.6	64	0.3
Shredded RDF produced in USA	11.2	26	27		
Eco-Briketts ex-German refuse	16	8–12	10–25	55–59	0.2–0.5
Typical UK industrial coal	30	6	6	34	1.6

Figure 5 shows that the corrosion rate is much reduced by avoiding the use of RDF on its own. There is a further reduction in corrosion rate if the boiler is operated at 260 °C compared with 482 °C, but of course this reduces the thermal efficiency of the plant.

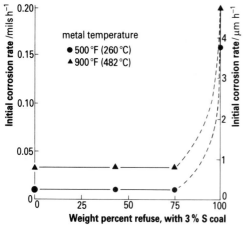

Figure 5
The initial corrosion rates (eight hour) for A106 carbon steel as a function of percentage of refuse by weight in fuel.
American Society of Mechanical Engineers

The flowchart for a pelleted RDF plant is like that in figure 3 (page 29), but with a pelleting press introduced after the fine shredder. Figure 6 shows the layout of the South Yorkshire County Council's installation which produces pellets for sale to industry, local authorities, and power stations. Figure 7 shows the operation process of the plant.

Figure 6
Layout of the South Yorkshire County Council's RDF installation at Doncaster.
South Yorkshire County Council

The pellets have a specific enthalpy (calorific value) of between 50 and 60% of that of industrial coal and sell for about 85% of the price of the equivalent quantity of energy from coal. The net cost of disposing of the refuse is about two thirds of that of long-haul transport and half that of simple incineration.

Characteristics of RDF

A comparison between various kinds of RDF and coal has already been given in table 2. The moisture content of the pellets influences their stability. Ideally the moisture content should not exceed 15%, and provided it does not exceed 20% pellets can be made which can be stored, under cover, for some months without marked deterioration. Since refuse can contain up to 30% of moisture (table 1) provision needs to be made in the pellet production process for drying the pellets if long-term storage is envisaged.

Conclusions

Production of RDF offers the possibility of a cost-effective solution to the disposal of municipal refuse in localities where environmentally acceptable refuse disposal sites are not immediately available and the alternatives involve long-haul transfer stations or the considerable capital and running costs of intensive incineration.

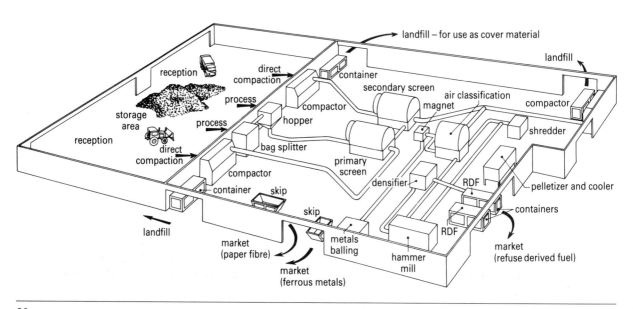

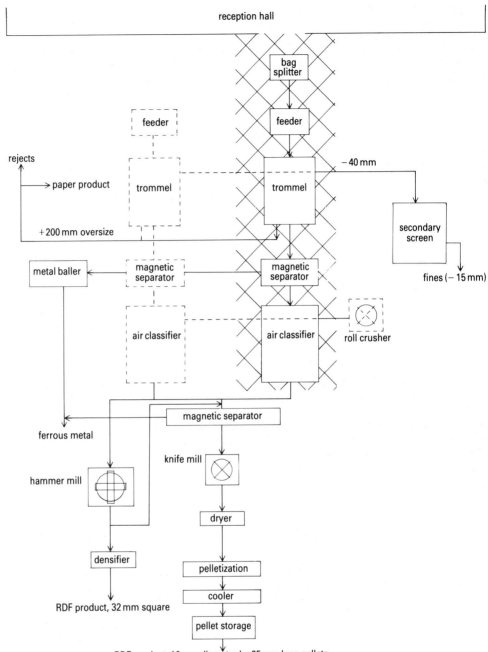

reception hall

bag splitter

feeder

feeder

trommel

trommel

−40 mm

secondary screen

fines (−15 mm)

rejects

paper product

+200 mm oversize

metal baller

magnetic separator

magnetic separator

air classifier

air classifier

roll crusher

magnetic separator

ferrous metal

knife mill

hammer mill

dryer

densifier

pelletization

RDF product, 32 mm square

cooler

pellet storage

RDF product, 16 mm diameter by 25 mm long pellets

Figure 7
Flowchart showing the operation process of the plant.
South Yorkshire County Council

The fuel can be cleanly and safely burned in conventional boiler plants, in mixes of up to 50 % on a heat-input basis, without significant boiler-tube corrosion. As fuel costs escalate, the prospects for RDF production will improve, and this form of resource conservation will help, in a small way, to conserve fuel supplies for future generations.

(This article has been adapted from one that first appeared in *Endeavour, New Series,* Volume **6**, Number 3, 1982. It is reproduced with kind permission.)

Brazil's Itaipú dam

JOEL FAGENBAUM
Technical Editor, *Mechanical Engineering*

At the end of 1983 the World's most powerful hydroelectric power generating complex, at Itaipú on the Paraná River between Paraguay and Brazil, went on line. Itaipú is expected to have a final installed capacity of 12 600 MW by 1988, and a capability of delivering some 79 000 GW h of energy annually. It will exceed the generating capacity of the US's Grand Coulee system, presently one of the most powerful hydroelectric plants in the World, by nearly 4 million kW. The generated electricity derived from the waters of the Paraná will be transmitted from this location to the industrial complexes of São Paulo, 600 miles to the east, and Rio de Janeiro, 200 miles farther. With over 30 000 Brazilians and Paraguayans working on the Itaipú dam alone, the estimated cost of this project could very well

Figure 1 (above)
Aerial view of the Itaipú dam during construction.
Carlos Freire, courtesy of the Alan Hutchison Picture Library

Figure 2 (right)
The construction of the Itaipú dam. An idea of the scale of the structure can be gained from looking at the size of the workmen.
Carlos Freire, courtesy of the Alan Hutchison Picture Library

approach $15 billion. Figure 1 is an aerial view of the site during construction, and figure 2 illustrates the enormous amount of work required in the building of such a project.

New high-power transmission links

When this massive hydroelectric complex is in full operation, Brazilian engineers will have to deal with the problem of transmitting very large amounts of power over long distances. At present, they are looking at the possibility of high-voltage a.c. transmission, 1000 kV or more, from the southeastern section of the country, where the load centres are concentrated, to the remote sites of the north, distances that approach 1555 miles.

Much of the country's electricity is obtained from hydroelectric generation, that is currently over 80% of Brazil's installed capacity of nearly 38 GW. Approximately one-half of this electricity is derived from the Amazon River, a quarter from rivers located in the southeast, west, and centre of the country, one-fifth from the southern regions, and the rest from Brazil's northeast. Itaipú will supply mostly the southern and southeastern areas of Brazil. This region is the most developed area of the country. It has the greatest concentration of industry and is responsible for three-quarters of the country's gross national product. A map of the area is given in figure 3.

By the end of the 1980s, Brazil is expected to put into operation over 1118 miles of high-voltage transmission lines, capable of handling from 345 to 750 kV. Most of this work will be done in the southern and southeastern part of the country. The estimated cost for the Itaipú transmission trunk alone is about $2.6 billion.

Power transmission: Paraguay and Brazil

All of the generated power from the Itaipú hydroelectric network will be fed into the Brazilian and the Paraguayan electric utility grids. Nine of Itaipú's generating units will operate at 60 Hz to match the Brazilian national electric system, while another nine will operate at 50 Hz to interface with the Paraguayan system. High-voltage (18 to 500 kV) step-up substations have been constructed with 50 Hz, 500 kV lines connecting the Itaipú powerhouse to a right bank substation. There are

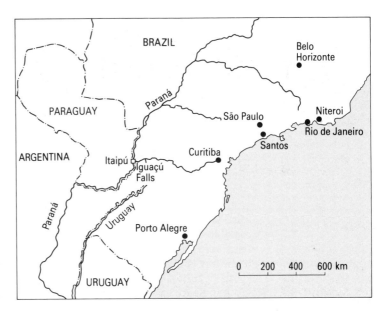

Figure 3
Map showing the southern part of Brazil, the Itaipú dam location, and the main industrial centres in which its power will be used.

also 60 Hz, 500 kV lines connecting the powerhouse to an a.c. step-up substation from which three 765 kV lines will transmit power to a terminal substation near São Paulo.

The Paraná spillway and hydraulics

The area of the Paraná River drainage basin and tributaries located above the Itaipú project site is 316 000 square miles. The land topography of the upper portion of the basin in the northeast is mountainous, the highest point being at 3280 feet. The northern half of the basin is tropical, with a rainy summer season and a dry winter, while in the southern half it is temperate, with hot summers and a nearly uniform rainfall throughout the year. The average annual rainfall in the basin above the Itaipú dam site is 55 inches, with precipitation ranging from a low of 47 inches in the upper part of the Rio Paranapanema basin, to a high of 94 inches in the Tietê and Chapin river basins.

Seven major tributaries feed the Paraná River before its confluence with the Igauçú River. The Iguaçú joins the Paraná about nine miles downstream of the Itaipú project site and it exerts a considerable influence on the Paraná's levels at the site.

Itaipú layout and site

The hydroelectric project area itself is located on the Paraná River about nine miles north of Ciudad Presidente Stroessner in Paraguay, and Foz do Iguaçú in Brazil. At this site, the Paraná flows almost due south and is about a quarter of a mile wide; at streambed level it is about 820 feet wide. From its streambed level the Paraná's two banks rise at an angle of almost 45° for a stretch of some 426 feet. Above that the terrain is quite flat, rising gradually from an elevation of 565 to 755 feet. The relatively flat plateau made it possible to construct highways on both banks of the river to provide access to the hydropower area from the cities to the south. A schematic diagram of the site is given in figure 4.

Diverting the Paraná River

The river diversion plan used for the Itaipú hydro project was based on various design approaches that engineers believed would be advantageous in light of the region's topographical features. Some of the principal ones were that rock from the diversion channel be used in the construction of the rock-fill dam, which had to be built concurrently with the excavation of the diversion channel; and the rock excavated for the foundations and approach channel of the spillway should supplement the rock from the diversion channel for the rock-fill of the main cofferdams.

In constructing the Paraná diversion control structure, builders wanted to provide a capability to handle a river flow of up to one million $ft^3 s^{-1}$ without causing

Figure 4
Schematic diagram of the Itaipú dam site.

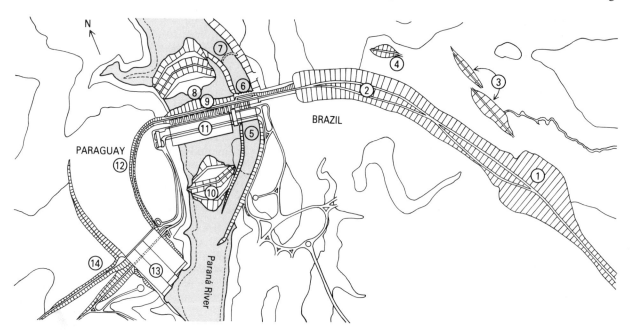

The structures include:

① left earth-fill dam, 7526 ft
② rock-fill dam, 6509 ft
③ Pomba-Qué dike and channel (a temporary structure)
④ Bela Vista dike (a temporary dam)

⑤ diversion channel
⑥ diversion control structure, 558 ft
⑦ arch cofferdams
⑧ upstream cofferdams
⑨ main dam, 3491 ft

⑩ downstream cofferdam
⑪ powerhouse
⑫ right wing dam, 3235 ft
⑬ spillway, 1279 ft
⑭ right earth-fill dam, 2861 ft

Other features are a spillway with 14 radial gates, 70 feet high and 65.6 feet wide, a hollow gravity-and-buttress version main dam with a 643 feet maximum height, and a solid gravity type and diversion control structure with a maximum height of 531 feet.

At full capacity, the plant will generate enough electricity to reduce Brazil's oil consumption by about 100 000 barrels a day.

any damage to the project's permanent works, and maximum flow of 1.25 million ft^3 s^{-1} without having the water overtop the main upstream cofferdam, at an elevation of 460 feet. The maximum average velocity of flow in unlined hydraulic passages of the diversion structure is limited to 49 ft s^{-1} for flow conditions of long duration. For short-duration flows, under some conditions, a 66 ft s^{-1} flow can be tolerated.

Machine layout in powerhouse

The final configuration selected for the Itaipú powerhouse consists of a set of 18 machine units, with 715 MW turbine outputs. A net design head of 388.5 feet permits a 5.25 feet loss at the intake and the penstock. The units will operate in such a way as to make the generating complex compatible with both the 50 Hz national Paraguayan utility system and the 60 Hz Brazilian electricity system. A sectional view of the powerhouse is given in figure 5.

Figure 5
A sectional view of the powerhouse.

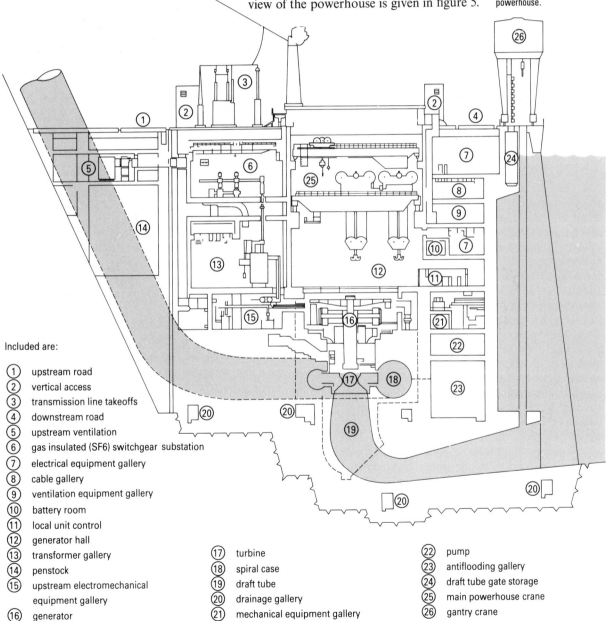

Included are:

1. upstream road
2. vertical access
3. transmission line takeoffs
4. downstream road
5. upstream ventilation
6. gas insulated (SF6) switchgear substation
7. electrical equipment gallery
8. cable gallery
9. ventilation equipment gallery
10. battery room
11. local unit control
12. generator hall
13. transformer gallery
14. penstock
15. upstream electromechanical equipment gallery
16. generator
17. turbine
18. spiral case
19. draft tube
20. drainage gallery
21. mechanical equipment gallery
22. pump
23. antiflooding gallery
24. draft tube gate storage
25. main powerhouse crane
26. gantry crane

An estimated maximum runner diameter for the turbines in the Itaipú powerhouse of 28 feet represented the largest size unit that could be transported from São Paulo to the Itaipú project site in one piece. This selection, Brazilian engineers indicate, would avoid the complications associated with split- and field-welded runner fabrications, or with split and bolted runners or completely site-welded runners. Figure 6 shows a turbine being fitted into place.

In designing the equipment for the power plant, Brazilian engineers had to consider the trade-offs involved between making the machine units as large as possible to reduce cost while maintaining plant reliability, state-of-the-art techniques from previous prototypes, and meeting transport limitations set on the shipment of heavy machinery.

Cooling: air-to-water heat exchangers

Due to the fact that power had to be supplied to both the Paraguayan and Brazilian national grids, operating at 50 Hz and 60 Hz respectively, two turbine systems had to be used. The Paraguayan turbines had to be able to operate for limited periods at gate openings greater than the rated 715 MW. The possible level of overload extends to the maximum limit of 740 MW imposed by the thermal restrictions of the generator.

To get over this problem direct water cooling of the stator conductors was selected to improve the reliability of this component. The purpose of this approach was to minimize the differential temperature caused by movements between the conductor and the slot iron in the stator windings.

Closed-circuit radial air cooling with air-to-water heat exchangers located around the stator frame was preferred for the generator rotors. Water cooling for the rotor was not selected because of the high inertia.

Power intakes and penstocks

There are 20 intakes to the Itaipú hydroelectric system. Sixteen are located in the hollow gravity blocks of the main dam and

four are in the diversion control structure.

All the intakes are similar in layout and design. Each one can pass $23\,308\,\text{ft}^3\,\text{s}^{-1}$ of flow with a velocity of $3.6\,\text{ft}\,\text{s}^{-1}$ at the trashworks, $53.8\,\text{ft}\,\text{s}^{-1}$ at the service gate, and $25\,\text{ft}\,\text{s}^{-1}$ in the penstocks.

The power intake structure, including the water passages, walls, piers, and beams, was designed and built of reinforced concrete, so that it would appear as a monolithic part of the block of the dam on which it is located.

Figure 6
A turbine being fitted.
Alan Hutchison Picture Library

Keeping debris out of the water intakes

Steel trashracks are used to keep debris from entering the intake system. There are 24 individual panels per intake arranged in four columns of six panels. Each panel is 15.2 feet wide and 18 feet high with a 6 inch clear space between the vertical bars.

The racks were designed for a normal differential head of 24.5 feet and a maximum differential of 33 feet. In addition, two trashrack cleaning machines will operate at the top of the dam (225 feet) to remove any accumulated debris. A head loss measuring system for the racks will activate alarms in the central control room when acceptable limits are exceeded.

Instrumentation to monitor structure safety

The principal concern of engineers regarding the safety of the Itaipú project is to monitor structural and foundational integrity. In the foundations, the parameters constantly measured are the uplift at the base of the structures and, in some parts within the foundation, seepage flowing through the foundation and the foundational displacements. For various project structures, selected parameters are under observation, such as absolute and relative movements, stresses and strains, opening of contractor joints (before grouting), and piezometric levels for earth- and rock-fill dams.

Brazil's hydropower future

Because of increases in basic industries such as petrochemical, heavy machinery, steel, and nonferrous metallurgy, Brazil has seen an average annual increase in the consumption of electrical energy during the last 30 years of about 10 %.

To meet this demand, new hydroelectric developments are under way in the planning stages even in the less developed areas such as in Tucurui, a jungle in northern Brazil. In the next few years a hydro plant with a 7000 MW capacity being developed there should be ready to start operation.

Over the next decade hydroelectric projects are scheduled for the Paranaiba basin, at Corumbá and Nove Ponte (500 MW plants), at Emborcaçao and Capim Banco (680 and 1000 MW plants), in the Paraná basin at Itha Grande, Porto Primavera, and Tres Irmãos (from 640 to 2000 MW plants), a 1400 MW plant on the Uruguay River, and a 1330 MW plant on the Araquaia River at São Felix. There will also be 160 MW thermal plants in the south and 1245 MW nuclear plants at Angra dos Reis near Rio de Janeiro.

(This article first appeared, in a slightly longer version, in *Mechanical Engineering*, November 1982. It is reproduced with kind permission.)

It is interesting to survey the wave-energy scene at this time since the research and development phase in the UK has recently been assessed by the government. The result was a decision to reduce effort on wave power in preference to other renewable technologies because of the disappointing cost estimates for electrical-energy generation by various notional schemes.

In this review we draw upon information gathered both from CEGB and Department of Energy studies. These involve all aspects of the system that is required to convert ocean waves into a marketable form of energy. Much of the publicity, however, has concentrated on the devices, and this aspect has stimulated inventions that number well into the hundreds. From these, about a dozen have deserved financial support for further investigation, and periodically attempts to reduce the number being studied have been made. This paper describes those services assessed in 1983, before the UK budget was significantly reduced.

The wealth of ideas for wave-power devices, whilst providing a stimulus, has created problems in deciding which concepts to pursue and which must be omitted from the research programme due to the finite resources of funding available. Other renewable technologies do not suffer from this problem to the same extent. With so many options for wave energy it is, therefore, not surprising that the research exercise has taken a long time. Figure 1 shows the various stages involved in evaluating a wave-energy system.

The subject matter of this article is in no way complete, and many problems, such as mooring and other major engineering features, are not discussed. However, the main aspects of the technology are reviewed, and particular emphasis is given to the fundamental problems. The reason for this approach is that new innovations are emerging which may change the detailed engineering aspects, and, in any case, it appears that a thorough understanding of the physics of wave-energy absorption could be the key to improvement.

Wave power: the story so far

B. M. COUNT, R. FRY, and J. H. HASKELL
Marchwood Engineering Laboratories

Wave characteristics

The first element of any system is, naturally, the waves themselves. Early power estimates were based on observational data from Ocean Weather Ship *India*, which suggested that an average power content of 90 kilowatts for every metre of wave frontage was available, ranging from periods with virtually no power to storm conditions when the power level is in excess of several megawatts per metre of frontage. More recent measurements at an inshore site more suitable for the development of wave energy devices at South Uist (see figure 2) demonstrated that average power levels in the range 40–50 kW m^{-1} are to be expected in water about 50 metres deep. Even further inshore, where the water depth is 20 metres or less and is

Figure 1
The requirements of a wave-energy power system.

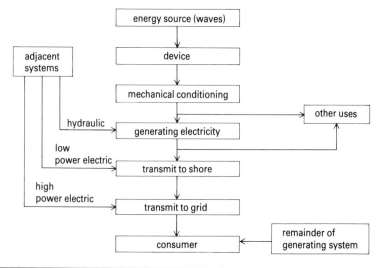

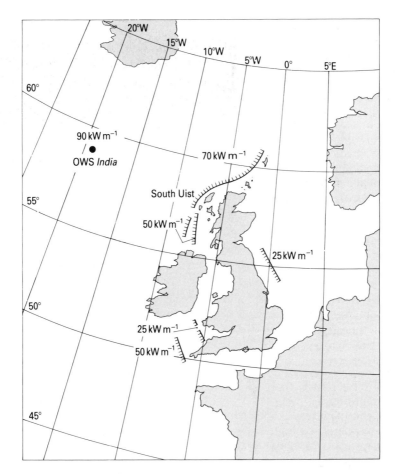

suitable to locate sea-bed mounted devices, the incident power level will be reduced, and there is much debate as to the rate of reduction.

It is quite surprising that after eight years of study there is still considerable debate on the wave climate, which after all is an essential input to any design or costing study. Not only are there considerable differences of opinion as to the rate of attenuation of power as the shoreline is approached, but the basic directional characteristics of the waves are not fully understood.

Ocean waves are, of course, generated by wind and are not simply monochromatic. Moreover, the ocean acts as a large integrator for wind energy. Waves transfer energy at an extremely high efficiency with attenuation distances of the order of several thousand kilometres – it is said that in the absence of wind the amplitude of a wave is halved in as many miles as the wavelength is measured in feet. It follows that waves measured at a particular location can either have been locally generated by the wind ('wind sea') or have originated in a storm a great distance away ('swell sea').

The Institute of Oceanographic Sciences and the National Maritime Institute have been involved in the collection and synthesis of wave data for many years and consequently have taken the leading role in this aspect of wave-energy research. Data collected by the IOS at South Uist over a period of two years has provided the most detailed description to date.

The method employed consisted of the measurement of wave elevation at a single point, and records over twenty-minute periods were collected every three hours over the two-year period. Each record was frequency analysed giving the form of the one-dimensional spectrum, and parametric analysis using spreading functions was used to determine the angular distribution of these waves. Results from this analysis are shown in figure 3.

In addition to this detailed work, meteorological models that can determine wave-power levels from wind data have been developed. These predictions have formed the basis of estimating the potential resource as 29 gigawatts if all suitable locations were to be exploited for wave energy. If one also assumes that 80 % of the locations could in fact be utilized and that the overall conversion factor to electricity is 30 %, a potential contribution of 60 terawatt hours per annum could be supplied to the CEGB, to be compared with the present annual electricity consumption in England and Wales of about 200 terawatt hours.

Capturing the energy

Hydrodynamic studies have shown that relatively small structures which are designed to move in response to the waves can be tuned to capture wave energy at a high efficiency. These devices can best be thought of as the ocean equivalent of radio antennae, and the major concepts that are being pursued all fall into this category, which will be divided into three subgroups: terminators, attenuators, and point absorbers.

Terminators

A terminator is defined as a wide structure which is aligned perpendicular to the incident wave direction. It is not surprising that this type of device is more fully understood than others at this stage of the study since early experimental work has been in narrow wave tanks, and theoretical calculations have been mostly two-dimensional. In both cases, devices of effectively infinite width are being simulated.

Of the seven devices assessed by the Department of Energy in 1983, four can be considered as terminators – the Bristol Cylinder, the Salter Duck, the NEL Breakwater, and the Sea Clam.

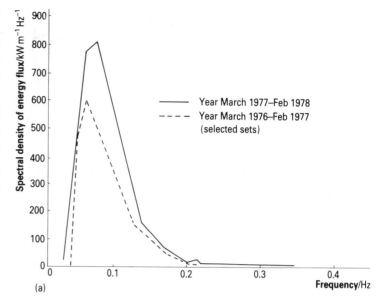

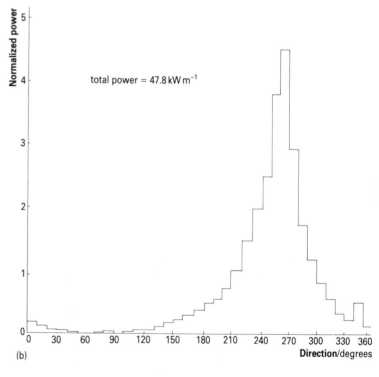

Figure 3
Results of wave energy and direction from a two-year study at South Uist.
(a) Average power spectra for offshore South Uist buoy.
(b) Mean distribution of power with direction.

The Bristol Cylinder

It has been observed that a submerged cylinder moving in a circular orbit would generate waves in a single preferred direction, and so from the time reversibility of the motion it is possible to absorb 100 % of the energy from a wave, if the motion of the device is controlled. This has been demonstrated experimentally for small-amplitude waves where control of both translation modes of motion was achieved, and now forms the basis of a study at Bristol University. An artist's impression of the device is shown in figure 4.

The potential advantage claimed for this device, apart from the high primary efficiency, is that it is fully submerged and therefore away from the violent interface during storm conditions. As a result of being submerged this does, however, mean that the sea bed is the only convenient datum against which the device may work and this may create mooring difficulties.

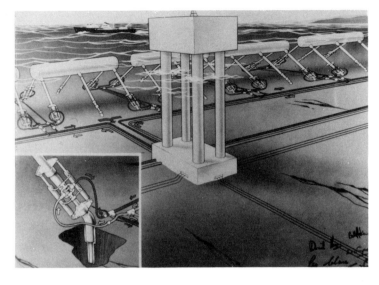

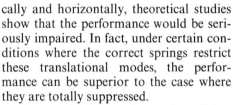

The Salter Duck

The 'Salter Duck' is an asymmetric cam-shaped device which is designed to extract energy through semirotary motion induced by incident waves. Whilst this device too could achieve 100% efficiency if all modes of motion were properly controlled, the practical option is to extract energy through the single rotary mode. As a consequence the performance will be less than ideal, although efficiencies in excess of 90% have been achieved in laboratory tests for monochromatic waves.

Since this device is intended to float, it is necessary to provide a working datum and restrict the translational motion of the device. The latter condition is necessary since, if it were totally free to move verti-

Figure 4
An artist's impression of the submerged-cylinder wave-power extraction device.

Figure 5
The 'Salter Duck' device, showing the use of gyroscopes inside the nose section.

cally and horizontally, theoretical studies show that the performance would be seriously impaired. In fact, under certain conditions where the correct springs restrict these translational modes, the performance can be superior to the case where they are totally suppressed.

As a result of these observations, Salter has proposed that his device should consist of a long central core, called a spine, upon which short duck sections are mounted. The virtue of the spine is that the angular spread of ocean waves over its length will introduce significant phase variations of the induced force and so the resultant will be low and the bulk motion small.

The duck must be regarded as extremely novel from an engineering viewpoint, and it is generally accepted that considerable development work would be required prior to any full-scale demonstration of this system. Perhaps the most unfamiliar concept is to propose that the gyroscopic power-conversion system will be housed in an evacuated pod and enough redundancy built in so that it will be maintenance free over a 25 year lifetime. The principle is illustrated in figure 5. Such a proposal is outside the practical capability of current heavy engineering and its feasibility could only be confirmed after some considerable research and development on various components.

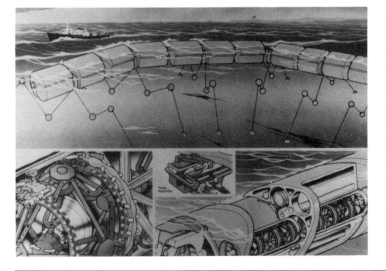

The NEL Breakwater

The oscillating water column is a device based upon the concept that an entrained column of water can be excited into motion by wave action and will therefore act as a massive piston capable of pumping large volumes of air. The low pressure air can be used to drive a turbine, thereby extracting energy from the sea.

A number of systems have been based on this concept and are being studied at the National Engineering Laboratory (NEL). The favoured system is a breakwater device, shown in figure 6, that is mounted on the sea bed in relatively shallow water. Unlike the Duck, proponents of this system claim that a full-scale demonstration could be achieved using current technology, and thus that it is the 'device of today'. However, there are widely differing opinions on the cost of providing what is, in effect, a breakwater, and there is almost a factor of two difference in the estimated cost of installation between different engineering consultants.

The Sea Clam

The final terminator device that is being supported in the national programme is the 'Clam', which is simply a flexible membrane that moves relative to some datum. As before, the datum may be supplied by a spine for floating systems or the sea bed for fixed ones.

The system is designed so that the space within the spine is sealed from the ocean and the trapped, pressurized air will be pumped by the membrane motion. This, in turn, is circulated around a high- and low-pressure pneumatic system and drives an air turbine. The system is shown in figure 7.

The major unknown with this system is the flexible rubber membrane, and extensive engineering studies on this aspect are still in progress. Another issue yet to be resolved is that the hydrodynamic performances of devices with a membrane at the fluid interface do not appear as good as other systems. It seems essential to determine whether this is a limiting feature or one that could be improved, and more work is necessary before a confident optimized design for this device could be undertaken.

Figure 6
The National Engineering Laboratory Breakwater device.

Figure 7
The 'Sea Clam' device. The hollow spine can be used to house electricity transmission equipment.

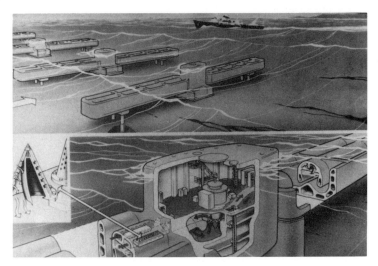

Figure 8
The 'Vickers Attenuator',
another oscillating water
column device.

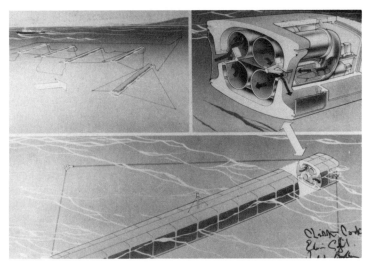

Figure 9
The Lancaster attenuator
device, an alternative to the
Sea Clam.

Attenuators

Attenuators are long, thin structures which are aligned parallel to the incident wave direction. The term attenuator came from the belief that energy would be progressively extracted along the length of the device, but recent studies have shown that for optimal operation the rear element of such a device would have to extract as much energy as the front one.

Only two attenuator devices were seriously studied in the United Kingdom and these are shown in figures 8 and 9.

The development of these systems may be limited, since CEGB studies demonstrate that optimized attenuators can only absorb about half as much energy as an equivalent terminator for a given volumetric displacement. As a result there must be doubts as to the potential of such devices.

Point absorbers

The final category of active devices are point absorbers, which are defined as structures that are small in comparison to the incident wavelength (say less than one-twentieth of a wavelength) and are arranged as arrays with no structural connection between adjacent devices.

It has been calculated that an axially symmetric device, constrained to move vertically, can absorb wave energy from an effective wave frontage of $\lambda/2\pi$ (λ is the incident wavelength of the wave). Moreover, when such devices are placed in an infinite line this absorption can be improved by as much as a factor of π. These are truly the equivalent of antennae arrays and can be arranged in a variety of different configurations.

In the UK an axisymmetric buoy enclosing a column is being developed at Queen's University, Belfast. The system can be either a moored or a floating device, or be rigidly connected to the sea bed. Its operation is the same as any other oscillating water column and therefore differs from the NEL concept only in its structural configuration and the ability to exploit the point absorber effects described above. Such a device is illustrated in figure 10.

The current assessment of this device, however, is not particularly favourable, not least because the hydrodynamic performances quoted above can generally only be obtained over a very limited wavelength bandwidth and require large amplitudes of movement.

The devices described above are a selection of the major concepts being investigated and other ideas are still being investigated in the UK and abroad. From this it is easy to see that there are many ways of converting wave energy to some intermediate form – mechanical, hydraulic, or pneumatic – at very high efficiencies. However, the problem of producing a marketable product from this primary form of energy has presented the greatest challenge to the various inventors and engineers.

Transmitting the power

Electrical generators driven from air or hydraulic turbines appear the best option for transmission of wave energy. Other energy-conversion possibilities include the use of geared or mechanically driven pumps to provide hydraulic power, simple brakes to provide heat, and the production of chemicals at sea.

In terms of cost, efficiency, convenience in use, and market value of the product that comes ashore, the direct transmission of electricity is the leading technology. This remains true even if one takes into account the life and reliability of the flexible cables that may be needed for some devices, the need to connect them to the sea bed, and all of the associated repair and maintenance costs.

The weakness of thermal transmission is that the market for a low-grade product – say water at 160 °C – at the remote sites where wave power would be landed is extremely limited. Moreover, transmission of hot fluid over large distances is prohibitively expensive. Whilst hydraulic transmission, using techniques developed for

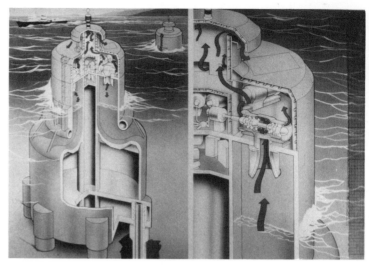

Figure 10
The buoy system developed at Queen's University, Belfast, which also uses the oscillating water column principle, but this time in the vertical mode.

the offshore oil industry, offers more hope, the technology is only viable when transmitting at high pressures. Moreover, the likely use of this high pressure fluid would be to drive turbines to produce electricity, and the hydaulic transmission pipe is both expensive and inefficient compared to submerged high-voltage electric cable.

When considering chemical fuels hydrogen is an obvious candidate, but even when produced from electrolytic processes with cheap electricity it is too costly compared to production using fossil fuels. Moreover, if electricity is produced for electrolysis it will surely be more sensible to use it directly, thereby saving fossil fuels for chemical uses. The same arguments, of course, apply to all hydrogen-based products.

The major engineering difficulties are at the primary interface, where the power conversion has to be versatile enough to provide the correct impedance to allow efficient hydrodynamic absorption, and still cope with the vagaries of the waves. Moreover, the ability to convert slow, irregular movements of the sea surface to fast, regular motion more familiar in electrical generation is a formidable task.

Typically, a 20 metre wide terminator section may transfer a few megawatts of power at a torque of 20 meganewton-metre and a rotational speed of 1.4 revolutions per minute (rpm). By comparison, a conventional 500 megawatt electrical generator rotates at 3000 rpm and requires only 1.6 meganewton-metre of torque. In the case of air buoys, a similar section would transfer its power through large volumes of air, say $300 \, \text{m}^3 \, \text{s}^{-1}$ at low pressures of about 0.1 bar above ambient. This power transfer is irregular on all time-scales with short term variations, where instantaneous power outputs ten times the mean level can occur.

Figure 11 shows the response of a typical wave-power device in an irregular sea state, and where linear velocity-proportional damping is assumed. The power output is then proportional to the square of the device velocity, and it is easily seen that there are instants when up to ten times the average power level is available. However, mechanical components are designed to transfer power up to a design rating, and it would not be sensible to rate the components highly, at great expense, in order to extract the peaks of the output, since in this case the machine would only operate at its design load for a small proportion of the time. On the other hand, if a smaller machine is installed, at a lower cost, there will be an associated loss of power transfer since the equipment cannot tolerate the power levels in excess of its rating. An economic compromise must be found.

In addition to this short-term variation, wave climatical conditions will vary throughout the year, with each condition having its second-to-second variability. Therefore, over the design life of any system – say 25 years – enormous waves may be expected, and calculations suggest that excursions of up to 15 times the root mean square can be expected. This, of course, will be a severe design problem, and is typical of those experienced in the offshore industry.

Whatever system is finally selected – gyroscopes, hydraulic pumps, or air turbines – it is easy to see that wave energy provides a challenge and is definitely not a re-appraisal of Victorian technology, as some people are led to believe. At present the air turbine offers a feasible route, but Salter believes that eventually the hydraulic routes will prove cheaper and more efficient, though they will need considerable further development.

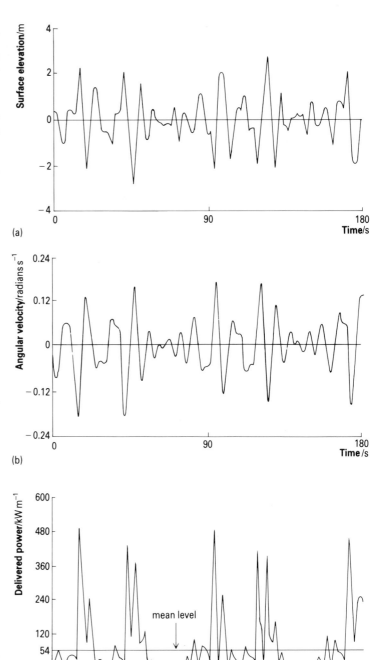

Figure 11
Predictions of typical responses of a linearly loaded wave-power device in an irregular sea.
(a) For a sea with energy of $80 \, \text{kW} \, \text{m}^{-1}$.
(b) Velocity profile of 16 m diameter duck with linear loading.
(c) Power delivered by duck with linear loading.

The future

The government announced on 2nd July 1985 that it had accepted ACORD's (Advisory Council on Research and Development in Fuel and Power) advice to discontinue work on wave power and geothermal aquifers in order to concentrate resources into more promising technologies. However, on wave power in particular, it was recognized that it will be important for the Department of Energy to be receptive to new ideas which might succeed in achieving the major cost reductions necessary to make it a worthwhile source of power.

Work is still progressing in other countries – notably Japan and Norway – and any future that wave power may have now rests on developments in these countries. In particular, the Norwegians are constructing two large demonstration devices, and the failure or success of this venture will determine to a large extent the level of worldwide support for wave power.

Acknowledgements

We should like to acknowledge all the device teams for assisting in providing information on their work, and the Energy Technology Support Unit for the artist's impressions used in this paper. In addition we thank the Institute of Oceanographic Sciences for permission to publish figures 2 and 3.

Further reading

CRABB, J. A., *Synthesis of a directional wave climate* (in *Power from Sea Waves*, ed. B. M. Count), Academic Press, 1980.

EVANS, D. V., JEFFREY, D. C., SALTER, S. H., and TAYLOR, J. R. M. (1979), 'Submerged cylinder wave-energy device – theory and experiment', *Applied Ocean Research*, **1**, 3.

Proceedings of second international symposium on wave energy utilisation, Trondheim, June 22–24, 1982 (ed. H. Berge), The Norwegian Institute of Technology, Trondheim.

WINTER, A. J. B. (1980), 'The UK wave-energy resource', *Nature*, **287**, 826.

(This article first appeared in *CEGB Research*, November 1983, and is reproduced with kind permission. It has since been slightly amended. All the illustrations are reproduced with kind permission of the Central Electricity Generating Board. When the article was published the government had just assessed the research and development potential of wave energy projects and had decided to reduce its financial assistance.)

Renewable energy: the Severn barrage

A. C. BAKER
Binnie and Partners, Consulting Engineers, London

In 1981 the report of the Severn Barrage Committee was published. This summarized the results of a study of tidal power which had lasted $2\frac{1}{2}$ years and was the most comprehensive report undertaken. In essence, the main conclusions were: that the construction of a Severn barrage would be technically feasible; that the best site was close to the Holm islands (figure 1), where a 7200 MW barrage, costing about £5.7 bn at December 1980 prices, would generate about 6% of the electricity presently supplied by the CEGB; and that the barrage would be an economic investment but less attractive than nuclear plant; and that no insurmountable environmental problems had been identified.

Origins of tidal power

It is well known that the tides are generated by the interaction of the gravitational fields of the Moon, the Sun and, to a lesser extent, the planets on the Earth, and that the time and height of high water

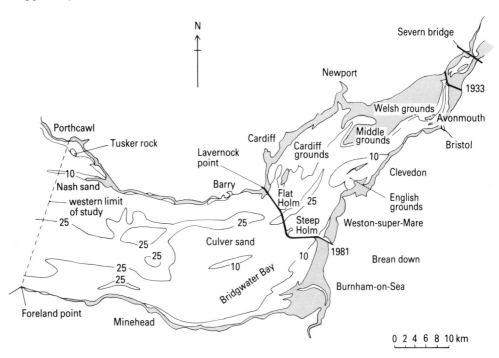

Figure 1
Locations of the barrages proposed in 1933 and 1981. The shaded areas are those exposed at low tide (above −5 m OD) and the contours are in m OD.

can be predicted accurately well into the future. This predictability is an important advantage of tidal power over wind or wave energy or, at ground level, solar energy.

In the open seas the tidal range is generally less than 1 m. Where geographical conditions are favourable, such as in trumpet-shaped estuaries like the Severn, the range at spring tides becomes amplified to 10m or more, resonance being an important factor. Once the spring tide range exceeds about 6 m then the prospects for tidal power begin to look interesting. In the Severn, equinoctial spring tides at Avonmouth can exceed 14 m, second only to the Bay of Fundy on the east coast of Nova Scotia, where a prototype Straflo turbine with a runner 7.6 m in diameter has recently been commissioned. This is the second largest tidal power scheme yet built, the largest being the 240 MW La Rance barrage near St Malo, France, which first produced electricity in 1966 (shown in figure 2). This points to another factor where tidal power is superior to the other electricity-producing renewables – it is well proven on a commercial scale.

A barrage for the Severn

The Severn barrage would have five main components: the turbogenerators; the power house structure (see figure 3); gated openings to refill the enclosed basin; a ship lock or locks; and embankments to fill the remaining gaps and provide a route for construction traffic and the power cables. Experience in the North Sea, the Netherlands, and elsewhere has shown that the large concrete (or perhaps steel) structures, called caissons, needed to house the turbines and the sluice gates could be prefabricated at suitable workyards and floated into position. This would be of great benefit to our steel and shipbuilding industries.

The last major construction work would be that of the embankments – composite structures of rock, sand, and intermediate gradings. Building these should be largely a marine operation, with the large quantities of rock being best brought from suitable quarries on the coast. Again, there

Figure 2
The 240 MW La Rance barrage, basin side.

is much relevant experience available.

During the studies carried out for the Severn Barrage Committee, much effort was put into identifying the most economic method of operating a tidal barrage, and then simulating the dynamic effects of water movement in the estuary and through the turbines or sluices. Figure 4 shows water levels over the spring tide cycle without a barrage and on both sides of the working barrage. The changes in water level in the basin when the turbines start up and close down can be clearly seen. These curves are for a barrage near the Holm islands, working as an ebb generation scheme (*i.e.* the discharge through the turbines is to seawards). Behind the

Figure 3
Cutaway view of concrete caisson for three bulb turbines.
Department of Energy and Binnie and Partners

Figure 4
Water levels both sides of a
barrage near the Holm
islands.
(a) At spring tide.
(b) At neap tide.
From One-dimensional model
studies of the Severn
Estuary, *Binnie and Partners,
October 1980*

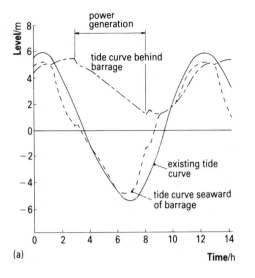

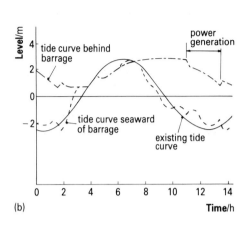

Figure 5
Typical output from a
Hydraulics Research Station
two-dimensional model,
showing the strength and
direction of currents with
and without the barrage.
*Hydraulics Research Station,
Wallingford*

barrage, a fundamental change in tidal regime will take place, low water being at about mean sea level, and high water, slightly lower than at present, lasting about three hours. The Severn bore would disappear.

Other methods of operation are possible, including flood generation, two-way generation (as at La Rance), and using the turbines as reversible pumps to raise the water level in the basin at high tide (also used at La Rance). In theory, by pumping at low head and using the water to generate electricity at high head, a net gain in energy results. Studies have shown that, after allowing for efficiency losses and extra capital cost, these options were less economic than simple ebb generation.

Figure 5 shows a typical result from a simulation model part way through the generating cycle.

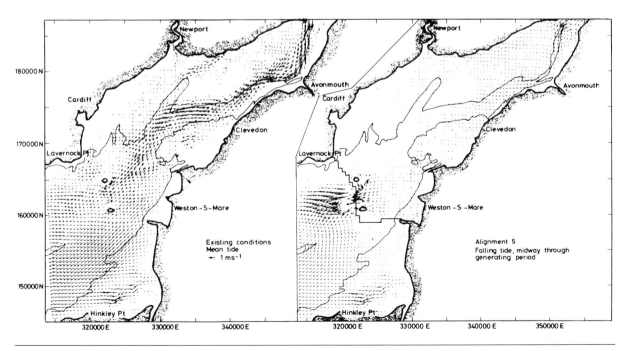

Affecting the environment

Space allows discussion of only a few of the many environmental factors involved if the Severn barrage were built. Although no insurmountable adverse effects have been identified, defining the 'acceptability' of the barrage would require much further study. The Severn Barrage Committee recommended that a comprehensive study, including preliminary engineering design work, should be undertaken immediately to allow government to take a decision on whether construction should proceed. This advice was not accepted.

In my view, the one environmental problem which may not be solved satisfactorily is the difficulty of the passage of smolts (young salmon) migrating to the sea. They are not fast enough to swim against the flow through the gates, and are unlikely to be willing to descend 15–20 m to pass through the turbines, so they may congregate in the relatively still water above the entrances to the turbine water passages. There they would be easy prey to predators. Catching them for transfer across the barrage or providing narrow fish ladders in the double walls of the caissons are possible solutions. If smolts do pass through the turbines, they are unlikely to collide with the blades, which would turn at only about 60 rpm, but they may be temporarily disorientated and thus again become subject to predation.

One benefit of the barrage would be the prevention of flooding caused by surge tides or high flows in the rivers draining into the basin. Ships sailing to the various estuary ports will make up delays in passing through the barrage locks by much easier passage through the port locks, because of the prolonged period of high water, higher minimum water level, and the reduced tidal currents. However, Sharpness, upstream of the Severn bridge, is the exception, and its lock would have to be altered to allow for generally reduced water levels.

Economics

There are two ways of looking at the economics of the Severn barrage: firstly, would money invested in its construction yield an acceptable return; and secondly, would the money be invested better in other electricity generating plant, such as nuclear plant?

The Severn Barrage Committee concluded that, using a real discount rate (that is, net rate above inflation) of 5%, the answer to both questions was generally 'yes', but depended greatly on what assumptions were made concerning the future cost of coal for power stations, on the amount of nuclear plant available, and its reliability (table 1).

It should be made clear that building the Severn barrage will not replace the nuclear programme. Development of the three best estuaries for tidal power (the Severn, Morecambe Bay, and Solway Firth) would supply about 12% of our present electricity requirements, with the useful feature that Morecambe Bay and Solway Firth have high tides about five hours after the Severn estuary, and so would generate their electricity 'out of phase'. Tidal power is the most economic, most proven, and most predictable of the UK's renewable energy resources for direct electricity generation, but a lot of nuclear plant will be required as well if the CEGB is to reduce its dependence on coal significantly. At least 2500 wind turbines of 90 m diameter would have to be sited on favourable coastal sites to match the output of the Severn barrage. The availability of this number of sites must be questionable.

Although the cost of building the Severn barrage would be huge, if spread over the proposed 12 year construction period it would require a rate of expenditure only about the same as British Rail needs for track maintenance. At a time of high unemployment in the construction industry, the real cost of the barrage would be much less than its apparent cost.

Table 1
Some published estimates of the economic worth of the Severn barrage

Source	Scenario	Severn barrage benefit/cost ratio*	PWR availability (as proportion of each year)	PWR benefit/cost ratio*
Report of the Severn Barrage Committee (EP46), table 13.7, 1981	High rate of increase in electricity demand, plus a high rate of nuclear development	1.0	70 %	1.99
	High rate of increase in electricity demand, plus a low rate of nuclear development	1.27	70 %	2.29
	As above, but lower PWR availability	1.28	60 %	1.87
Strategic Review of the Renewable Energy Technologies (ETSU R13), 1982	Low rate of increase in electricity demand, plus a medium rate of nuclear development	2.8–3.4†		
	Medium rate of increase in electricity demand, plus a medium rate of nuclear development	1.6–1.9		
	High rate of increase in electricity demand, plus a medium rate of nuclear development	1.2–1.4		

* Benefit/cost ratio is measured by assembling the net lifetime benefits, *i.e.* the value of fuels and generating plant saved each year less the cost of operation, and discounting these to a chosen date at 5 %. This figure is divided by the costs of construction similarly discounted. A benefit/cost ratio of 1 indicates that the investment will show a rate of return of 5 % above inflation.

† The lower figures in this column are based on barrage costs similar to those used by the Severn Barrage Committee. These include all design and development costs and an allowance of about 30 % for cost overrun excluding inflation. The higher figures assume only 10 % cost overrun and slightly lower turbine costs. The overall conclusion is that the economics of the Severn barrage are broadly comparable with those of nuclear plant.

Further reading

HEADLAND, H. (1949), 'Tidal power and the Severn barrage', *Proc. IEE*, **96**, Part 2, No. 51.
Strategic Review of the Renewable Energy Technologies: an economic assessment. Energy Technology Support Unit, Report ETSU R13, November 1982, HMSO.
Tidal power from the Severn Estuary, Energy Paper No. 46, 1981, HMSO.

(This article first appeared, in a longer version, in *Physics Bulletin*, Volume **35**, 1984. It is reproduced with kind permission.)

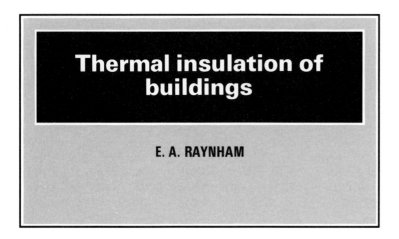

Thermal insulation of buildings

E. A. RAYNHAM

As energy becomes more and more expensive, so it becomes necessary to use it more sparingly and to conserve it more carefully. But even today as the media are screaming 'energy crisis' and fuel bills rocket, householders still use (and therefore pay for) more fuel than they need to warm their homes. This is despite government campaigns such as 'Save it', which encourage us to keep the heat we have paid for so dearly.

Improved thermal insulation of buildings is directed at reducing the overall rate of flow of heat from them, thereby conserving fuel in cold weather and securing better comfort when ambient temperatures are high. It is also directed at improving comfort at reduced cost by reducing draughts which may originate from admission of cold air through direct openings to the outside or from local regions of high thermal loss, such as windows.

Causes of heat loss from buildings

A hermetically sealed building is ideal from a heat loss point of view. Clearly if no cold air is coming in, being heated, and then flowing out, one source of heat loss is eliminated. However, this is not a practical proposition, since sufficient air must be allowed to flow through a building to permit the necessary combustion processes to proceed with safety and comfort. These are the burning of fuel for cooking, heating, to a lesser extent lighting, and, most important of all, breathing. The guide book (1970) of the Institution of Heating and Ventilating Engineers (IHVE) suggests, for example, that half an air change an hour is necessary for bedrooms, one per hour for living rooms, one and a half per hour for lavatories, two for bathrooms, and five for cells in police stations!

In fact, one and a half air changes per hour is considered adequate for most situations in dwellings, offices, shops, factories, hospitals, and the like.

Excessive ventilation

Tests conducted some years ago by the Building Research Station on brick built houses with standard factory-made joinery indicated that they suffered some two and a half air changes per hour. Such a level of ventilation would result in a heat loss from a pair of typical semidetached houses of approximately $470 \, \text{W K}^{-1}$ temperature difference between the air inside and outside the dwellings. This is half as much again as conducted heat loss through cavity brick walls of the same two dwellings, and nearly twice as much as the heat loss through their uninsulated roofs. Clearly then, to reduce heat loss from buildings the first priority is to control ventilation. This can be done by weather stripping doors and windows, either with metal strips or with self-adhesive plastic foam. These measures are not perfect and are most unlikely to reduce the air change below the one and a half referred to above as the acceptable minimum.

Windows

It would perhaps be convenient here to compare heat lost by ventilation with conducted heat losses attributable to windows. It is commonly assumed that double glazing reduces heat loss through windows by 50 %. In fact, the IHVE guide book shows that the conduction heat loss through a double-glazed wooden window is $2.5\,\mathrm{W\,m^{-2}\,K^{-1}}$ compared with $4.3\,\mathrm{W\,m^{-2}\,K^{-1}}$ through a similar single-glazed window. The corresponding figures for metal windows are 3.2 and $5.6\,\mathrm{W\,m^{-2}\,K^{-1}}$ respectively.

It should also be realized that, in the case of the pair of houses already referred to above, conduction heat losses through windows amount to some $130\,\mathrm{W\,K^{-1}}$ temperature difference between inside and outside air. Double glazing would reduce this by not much more than $50\,\mathrm{W\,K^{-1}}$.

Two Building Research Station reports (D. A. Thomas and J. B. Dick, 1953, and J. B. Dick, 1950) permit a comparison to be made between the effects of weather stripping and double glazing. This comparison is summarized in figure 1 and shows that weather stripping does far more to reduce heat loss than does double glazing. Indeed, unless double glazing is fitted in such a manner as to reduce ventilation losses as well, it will not be effective or economical. It must be remembered, however, that double glazing does provide other benefits, such as reducing down draughts, reducing condensation, and improving security.

Walls and roofs

Any material or technique used to reduce heat flow through walls or roofs does so by incorporating air or other poorly conducting gases into the structure. This is often done by using lightweight materials which themselves contain trapped air. The lightweight materials that are available include mineral wool, quilts and slabs, expanded polystyrene slabs, foamed polyurethane slabs, low-density wood-fibre insulating boards, cement-bonded slabs, and strawboard. Heavier materials, which also make a contribution to thermal insulation, include lightweight aggregate concrete blocks and aerated concrete blocks. Apart from the use of such materials, air can also be incorporated into a structure by forming a cavity, such as the conventional cavity wall used for house building. The insulation of such walls may be further improved by injecting foamed plastic cavity fills between the walls.

Whilst some of the above materials are best installed during construction (lightweight concrete blocks for example), others may be added at any time (mineral wool quilts in loft spaces, providing the loft is accessible). Fortunately, there are other techniques and materials which may be used to effect thermal insulation improvements to existing walls and roofs. An additional wall lining fixed to timber battens creates an insulating cavity. Moreover, since half the heat flowing across such cavities is radiated, the insulation performance of the cavity may be almost doubled by including a low-emissivity surface in the cavity. This is done, for example, by using aluminium foil backed plasterboard. An alternative, used by the author in his dining room, is to fix the foil under the battens and then use a predecorated rigid form of medium density hardboard.

Since the effectiveness of the above materials is dependent on thickness it cannot be compared with, say, the effectiveness of a cavity by referring to thermal conductivity figures. It is usual, therefore, to use thermal resistance as a parameter. This is the product of thermal resistivity, k^{-1} (where k is the coefficient of thermal conductivity), and thickness, and is expressed as $\mathrm{m^2\,K\,W^{-1}}$.

Figure 1
Comparison between conduction and air infiltration heat losses through windows. *a* is air infiltration through a weather-stripped window, and *b* is the equivalent heat loss. *c* is heat loss by conduction through a double-glazed metal window. *d* is air infiltration through a window without weather stripping, and *f* is equivalent heat loss. *e* is the heat loss by conduction through a single-glazed metal window.

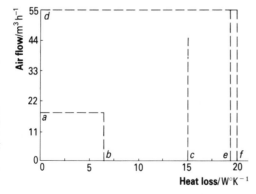

Shorter heating seasons

It is possible to use such large amounts of thermal insulation that the heat created within a building may almost suffice for maintaining comfortable temperatures, although this may not be a very practical solution. However, it has been pointed out on a number of occasions that provision of adequate insulation makes heating unnecessary until the daily mean outdoor temperature falls below about 14 °C. In other words, the heating season for most of the UK could be reduced by almost two months, to run from late October to mid April, instead of from early October to mid May.

Fuel costs

The sensible level of expenditure on insulation is directly related to the cost of the heat lost, and to the length of time required to produce an appreciable return in terms of fuel saving. Since fuel costs differ so widely, it is not possible to generalize on amounts of insulation which should be used. Until recently it has been thought reasonable to assume a two or three year period for the insulation cost to be compensated by the accumulated fuel bill reductions. However since fuel costs are likely to go on rising in the foreseeable future it would seem prudent to install more than this amount of insulation.

Condensation

To avoid condensation on the surface of walls, windows, and so on, it is necessary to have sufficient ventilation, and adequate insulation and heat input. It is also necessary to consider the question of whether condensation is likely to occur within a structure. Since most structures experience a falling dew point as well as temperature gradient from inside to out, it is possible (especially where insulating internal linings are provided) for the temperature on the cold side of the insulation to fall below the dew point, causing interstitial condensation. This can be controlled by providing water vapour barriers on the warm side of the insulation. This may take the form of polythene film or certain types of paint treatment, such as chlorinated rubber, on the insulated lining. In some situations, such as factory roofs and some timber flat roofs, ventilation is provided above the insulation to remove any water vapour that has penetrated that far.

Conclusions

It can be seen that clear benefits accrue from insulating existing buildings to a higher standard, and that when doing so the first priority should be reduction of adventitious ventilation. The insulation techniques and materials required are readily available and easy to apply.

In new constructions the prospect of improved standards is welcome. The scale of the improvement is such as to call for new techniques and combinations of materials. The result may indeed be such as to have a profound effect on building design and construction generally.

The matter will need constant review as fuel costs change. Moreover, if the maximum benefit of all these efforts is to be achieved, a more sophisticated approach to heat loss calculation and control is required.

Finally, the problems of interstitial condensation which may result from improvements in thermal insulation must not be ignored.

Further reading

DICK, J. B. (1950), *IHVE Journal*, **18**, 17–9, pp. 123–34.

IHVE Journal (1946), **14**, 135, pp. 103–12.

IHVE Guide Book (1970).

IHVE Journal (1974), **42**, 63.

RAYNHAM, E. A. (1971), *Proc. Nat. Symp. on Structural Insulation*. London: Structural Insulation Association.

RAYNHAM, E. A. (1973), *Fibre Building Board Development Organisation Technical Bulletin* TB, pp. 10–73.

THOMAS, D.A. and DICK, J.B. (1953), *IHVE Journal*, **21**, 214, pp. 85–97.

(This article first appeared, in a slightly longer version, in *Physics in Technology*, July 1975. It is reproduced with kind permission.)

Introduction

Wind energy use has several attractive features. Usually high winds occur in areas with low land values, such as moorland and mountain regions; little area is taken up by the windmills or other collecting equipment; collection does not greatly disrupt activities in the surrounding area; and wind energy provides a non-polluting and abundant renewable resource – a possible 100 million million kW h (kilowatt hours) per year in the US alone.

The main problems in the large scale use of wind stem from its great variability and the capital intensiveness of the necessary machinery. Although wind speeds are stable in a given place on an annual basis they are very variable on hourly, diurnal, and seasonal bases.

Since wind power varies as the cube of the wind speed, this variability is a very important consideration when looking at the potential of wind as an energy source. It means that an energy storage and reconversion system is required to smooth out the energy supply.

This article attempts to present an up-to-date review of various energy storage and reconversion options that are possible and/or have been proposed for use with wind energy conversion systems.

Energy storage options

Wind energy can be easily stored in the four ways noted below:

1 Mechanical energy storage: as potential energy (*e.g.* in conjunction with hydroelectric systems), or as kinetic energy in flywheels.
2 Chemical energy: as hydrogen, synthetic fuels, and fertilizer, or in secondary batteries.
3 Thermal energy: as sensible heat, as latent heat, or in thermochemical storage in reversible chemical reactions.
4 Electric or magnetic energy: in high-intensity electric or magnetic fields.

Storage options for harnessing wind energy

R. RAMAKUMAR
Oklahoma State University, Stillwater, Oklahoma

Mechanical storage

Hydrofirming

Hydrofirming involves linking a wind energy system to a, generally larger, hydroelectric system. The wind energy can be used to provide energy to pump water into a storage reservoir. In essence, wind energy is stored as potential energy in water.

Linking wind and water energy systems benefits from their complementary peaks in productive power. More water is frequently available in the period of spring run-off, whereas wind speeds are usually greatest in the winter. The extra wind energy can be used to supplement a hydroelectric system at times of peak demand particularly, and, more generally, to help to smooth out the annual energy production. The optimum scale of wind energy production will depend on the scale and characteristics of the hydroelectric system with which it is being used. The wind energy can also be used to replace depletable energy resources such as coal, oil, and natural gas in the running of the power station. There are, however, problems relating to the oscillatory stability of the aeroturbines, machine interactions between the two energy production systems, and restrictions on the machinery posed by the minute-to-minute wind speed fluctuations.

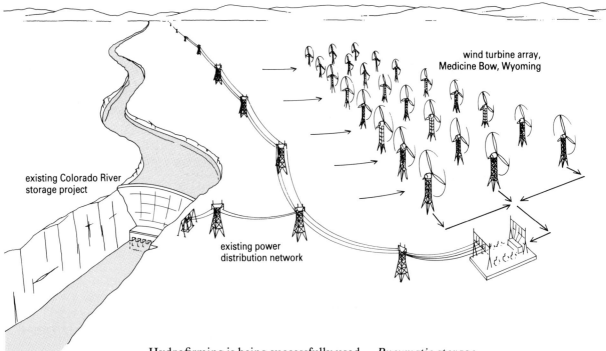

Figure 1
Schematic diagram of the
Medicine Bow hydrofirming
plant.

Hydrofirming is being successfully used at the Medicine Bow hydroelectric plant in Wyoming, USA, where a major effort was made in 1977 to integrate 100 MW of wind generation with the existing 1324 MW hydroelectric capacity. A schematic diagram of the system is shown in figure 1.

On a smaller scale, aeroturbines operating in limited (<50 kW) wind energy systems can be used to operate pumping wells to store water for future electricity production in individual residential installations.

Pneumatic storage

Wind energy can be used to operate a multistage compressor to compress air for storage in underground caverns or overhead tanks (see figure 2). Overall efficiencies of energy conversion in this manner are low (only about 30 %).

This efficiency can be substantially increased by expanding the compressed air through a combustor and turbine driving a generator (figure 3). Such systems, with fuel assist (inputs from outside), can have efficiencies of the order of 45 %. They require

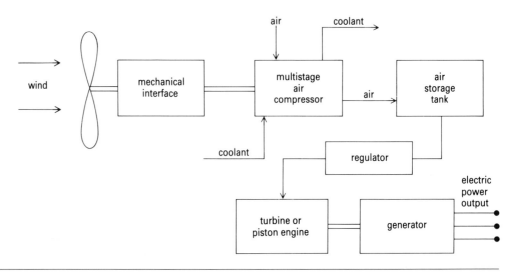

Figure 2
Pneumatic storage of wind energy.

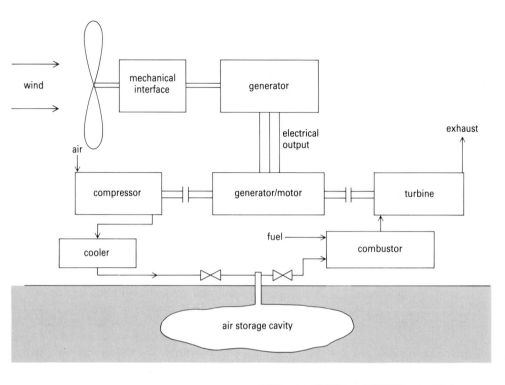

Figure 3
Compressed air storage of
wind energy.

very small areas above ground and use
existing turbine technology.

Flywheel storage
Wind energy can be stored as kinetic
energy in a flywheel, as shown in figure 4.
The energy stored is proportional to the
square of the rotational speed of the flywheel, giving rise to 75 % conversion efficiencies. Various materials for making flywheels have been suggested, ranging from
maraging steel to composites of epoxy and
glass. Typical energy storage parameters
are given in table 1.

Table 1
Energy storage parameters for flywheel materials
(based on a speed ratio of 3:1)

Material	$W h kg^{-1}$	$W h l^{-1}$
Plywood	4.2–5	2.9–3.4
Glass	9.7–16	11.4–18
Kevlar	50	37
Maraging steel	55	
Advanced anisotropic materials (projected)	220	

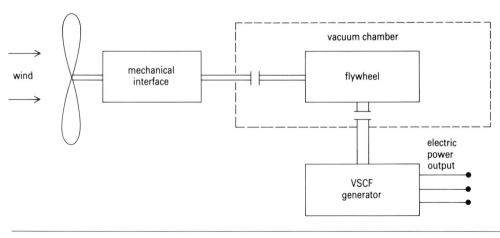

Figure 4
Wind energy storage in
flywheels.

Prototype flywheel systems have shown specific energy densities in the range of 25 to 50 W h kg^{-1}, with predictions of up to 220 W h kg^{-1} in the future. Although promising, flywheels are, at present, suitable only for short-term buffering storage. Work is needed to bring down their cost and improve their reliability. The overall efficiencies of mechanical systems are shown in table 2.

Table 2

Typical round-trip (electrical-to-electrical) energy efficiencies of mechanical energy storage systems

Technology	Efficiency range (%)
Potential energy of water (pumped storage)	65–75
Compressed air energy storage (without fuel assist)	25–35
Compressed air energy storage (with fuel assist; overall)	40–45
Flywheel energy storage	75–85

Chemical storage

Hydrogen storage

Wind-induced energy can be used to dissociate water into hydrogen and oxygen molecules by electrolysis. The hydrogen can then be stored for future use as fuel or chemical feedstock.

Production of hydrogen by electrolysis can be made up to 90 % efficient by operating cells at high pressures (6 to 20 MPa) and at temperatures of 150 to 200 °C.

Hydrogen can be stored in three ways: as a high-pressure gas in cylindrical steel storage vessels, as liquid hydrogen, or as metal hydrides. Metal hydrides appear the most promising as the other methods have problems of storage area and continuous refrigeration respectively. However, at present all three methods are very capital intensive.

Synthetic fuels and fertilizer

Problems relating to hydrogen storage can be mitigated by synthesizing an array of hydrocarbons, alcohols, anhydrous ammonia, or ammonium nitrate for easy and inexpensive storage. Nitrogenous fertilizers are synthesized from nitrogen (from the air), hydrogen (electrolysed by wind energy), and feed water.

The 1978 Lockheed study concluded that wind systems sized to produce anything more than four tonnes per day could economically compete with conventional plants producing 100 tonnes per day, a great boon to localized production of fertilizer.

Electrochemical storage

Electrical energy can be stored in 'secondary batteries' by chemical energy changes. Lead-acid batteries run off wind chargers were in use in rural areas before electricity supply extended to these places, and could be used again.

Battery energy storage is efficient, silent, and virtually pollution free. It can be used for levelling energy loads, powering vehicles, and in aerospace projects. As work on secondary batteries develops, so the potential of wind-powered battery plants will increase.

Thermal storage

Wind energy can be converted and stored as thermal energy in a variety of ways, as illustrated in figure 5. A mechanical water churn can be used to heat water for storage in low-grade (100 °C) thermal form. A resistance heating element supplied by the unregulated (variable voltage, variable frequency) output of a wind-driven alternator is another possibility. The use of suitable capacitors can allow very good matching between the power in the wind and the power output of the system. Thermal storage has great potential for providing energy for family residences, especially in farms for such uses as house heating and crop drying.

Sensible heat storage

When the temperature of a material is raised, a certain amount of energy is stored as sensible heat. This is by far the easiest and most economical way to store thermal energy. Materials such as steel and sand can store energy at up to 400 °C, whereas storage in water is limited to 100 °C. The expense of storage materials can be reduced by using dual medium storage

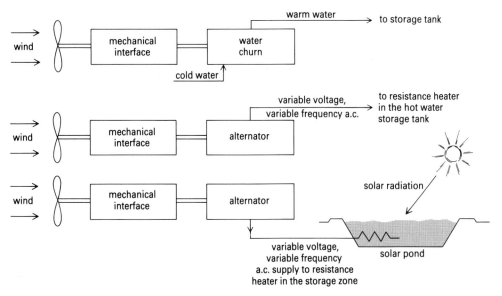

Figure 5
Techniques for storing wind
energy in thermal form.

materials, such as stratified (salt-gradient) solar ponds which can be used to store both wind and solar energy (figure 5). Such a system can take advantage of the complementary nature of wind and solar energy production.

Latent heat storage
Low cost materials such as salt hydrates, ice, and wax can be used to store energy in their isothermal changes of state (from solid to liquid and liquid to gas) at temperatures around 150 °C. Sodium sulphate decahydrate (Glauber's salt) has received special attention because of its low cost and relatively high latent heat of fusion ($255 \, \text{kJ kg}^{-1}$). For storage at higher temperatures (300 to 800 °C) several nitrates and fluorides have been suggested. Energy storage parameters for sensible and latent heat storage materials are given in tables 3 and 4.

Sensible and latent heats can be combined using large temperature ranges covering changes of state, the total energy stored being given by the simple addition of sensible and latent heats.

Electric or magnetic storage

Energy stored in electric and magnetic fields is proportional to the square of the field intensity. Although high intensity fields have great potential for storing energy, the technology available allows for only short duration storage, for instance with the use of capacitors. To create the high intensity magnetic fields, superconducting magnets are required, which, although available, are still very expensive. They are economical only when very large and thus are incompatible with small, individual wind energy units.

Table 3		
Selected sensible heat storage material parameters		

Storage medium	$\text{W h kg}^{-1} \, {}^\circ\text{C}^{-1}$	$\text{W h l}^{-1} \, {}^\circ\text{C}^{-1}$
Solids		
Granite rocks	0.244	0.644
Concrete	0.314	0.703
Cast iron	0.233	1.839
Liquids		
Water	1.167	1.111
Oils	0.43–0.76	0.53–0.73
Sodium	0.35	0.336

Table 4				
Selected latent heat storage material parameters				

	Heat of fusion			Melt temperature/ °C
	W h kg^{-1}	W h l^{-1}		
Material		Solid	Liquid	
Glauber's salt	62.5	91.3	83.1	32
Paraffin wax	58.1	47.6	44.7	47
Sodium nitrate	50.3	113.6	95.6	307
Mixed fluorides	170.8	439.0	357.0	832

Conclusion

Although wind energy systems can only be expected to have capacities of a few tens of MW, small compared to other systems with 500 to 1000 MW capacity, they do have great potential both when integrated with existing systems and as small, individual energy producers. Wind energy is likely to be linked to the general electricity system and dispersed throughout the system, giving rise to the term 'dispersed storage and generation' systems often associated with wind-energy production.

Effective integration still faces problems of the limitations of the machinery under extreme wind conditions and how best to link wind systems to the existing energy production systems. These are, however, being tackled.

This article has looked at the available storage options for wind energy, and, while there is obviously huge potential, a great deal of work has to be done on development of the necessary equipment. It is hoped that in the near future wind energy will be able to play an important role in meeting overall energy requirements.

(This article is amended from one that first appeared in *Mechanical Engineering*, November 1983. It is reproduced with kind permission.)